The Whole Story

The Whole Story

Alternative medicine on trial?

Toby Murcott

Macmillan

First published 2005 by
Macmillan
Houndmills, Basingstoke, Hampshire RG21 6XS and
175 Fifth Avenue, New York, N. Y. 10010
Companies and representatives throughout the world

ISBN-13: 978–1–4039–4500–6
ISBN-10: 1–4039–4500–4

This book is printed on paper suitable for recycling and made from fully managed and sustained forest sources.

A catalogue record for this book is available from the British Library.

A catalog record for this book is available from the Library of Congress.

10 9 8 7 6 5 4 3 2 1
14 13 12 11 10 09 08 07 06 05

Printed and bound in China

Dedicated to the memory of my father,
Ken Murcott (1939–2004)

Contents

acknowledgements

This book needed to feature a wide range of topics and ideas in which I can claim no expertise. Therefore I owe a great many thanks to the specialists who have been so generous with their time and thoughts. When I have got it right, it is they who have made it so. Any remaining mistakes are mine and mine alone.

I have had numerous conversations with many people, either by phone or email, all of whom have contributed in some way – sometimes explicitly as acknowledged in the text, sometimes by keeping me on the right track. For these conversations I would like to thank Bernadette Carter, Iain Chalmers, Sabine Clark, Rachel Clarke, Mike Cummings, Robert Dingwall, Edzard Ernst, Jane Gallagher, Jenny Gordon, Gill Hudson, Janice Kiecolt-Glaser, Kate Kuhn, Catherine Law, Richard Nahin, Robin Lovell-Badge, Alan Marsh, Hannah Mackay, Andrew Moore, Stephen Myers, Ton Nicolai, David Peters, Paolo Roberti, Virginia Sanders, Aslak Steinsbekk, Andrew Vickers, Harald Walach and Graham Ward. I would like to extend an additional thank you to John Hughes for permission to describe his as yet unpublished data and to Paul Drew, Louise Fletcher and Virginia Olesen for additional help. I am also very grateful to Christine Barry, John Chatwin and Zelda Di Blasi for permission to draw on material from their PhD theses. I also owe a special debt of gratitude to Hilly Janes for allowing me the opportunity to explore some of these ideas in print over the last year.

There are two people who have been a particular help to me both by being generous with their ideas and with their contacts. Paul Dieppe and George Lewith, in different ways, gave me keys to the world of complementary medicine research.

The support and advice of my editor, Sara Abdulla, as well as her efforts on the manuscript, have been invaluable. It's a testament to her courage, or perhaps foolhardiness, that she was prepared to take on a novice author, and she has earned my considerable gratitude as a result.

The process of writing the book would have been impossible without the tolerance of my friends and family who have accepted my constant refrain of 'when the book is finished' with great good humour. Two people in particular have helped me far more than I can possibly express. Anne Murcott patiently guided me through a crash course in sociology and put her academic experience at my disposal: as she insists on putting it, it has been quite handy that she is also my mother. Above all, I am especially grateful to Kerry Chester, my partner and companion, for generously allowing herself to take second place to my writing and for putting up with a one-track monster in the house.

preface

I spent eight years as a research biochemist. Throughout that time I lived with Sam, a large, ginger neutered tom cat. He had moved in with no fur on his belly and back legs – apparently as a result of fleas – and a medicine cabinet of powerful and expensive steroids. I decided that if fleas were causing Sam's baldness then I would deal with them rather than spend my meagre funds on cat steroids. So I bought him a flea collar and dusted him down with flea powder. Before long his fur had grown back.

Ten years later Sam became lethargic and lost interest in his food. The vet diagnosed kidney failure and gave him a few weeks, or at most a couple of months, to live. I'd grown very fond of Sam, so despite it being against my better, scientifically trained, judgement, I followed a friend's advice and took him to a homeopathic vet.

Slightly to my surprise, the homeopathic vet gave the cat a cursory once-over then spent the next twenty minutes grilling *me*. Was he good natured or grumpy? Did he like warm spots to sleep in? How did he eat his food? In the end he announced that Sam was an angry cat and that, as it often did in cats, his anger had settled on his kidneys – and by the way had he ever lost his fur? When I said that he had, the vet replied that it was a common reaction in angry cats. I kept my thoughts to myself, knowing that the best explanation for his fur dropping out was a flea allergy and that removing the fleas cured the problem. Angry kidneys had not till then figured anywhere in my thinking, let alone my biochemical training. But having got this far, and confident that the pills would do no harm – I assumed them to be sugar pills – I duly gave Sam the homeopathic remedies prescribed.

The vet's parting shot had been to warn me that his fur might drop out again and not to worry if it did. That would be the remedy drawing out the anger from his kidneys and making him better. A week later we started to notice clumps of ginger fur on the carpet, and sure enough, the baldness had returned in exactly the same pattern as the earlier fur loss. It soon grew back and Sam went on to live for another year, beating the conventional vet's prognosis by a considerable margin.

Was this my damascene moment? Did I convert to the faith of homeopathy and abandon my scientific career? Not a bit of it. My scepticism remained – remains – intact, yet my scientific training had taught me not to dismiss uncomfortable observations out of hand. It started me thinking: how could it be explained?

There are at least four possible explanations for what happened to Sam: therapeutic effect; coincidence; placebo; and conjuring trick. Working through each alternative rapidly became less interesting than considering the grounds for deciding which was the most reliable explanation. Rather than ask whether or not such therapies work, the first questions are:

How can we tell if they work? What methods do we have that will tell us? This book is the result of pursuing these questions.

Toby Murcott
Bristol, Spring 2004

1
introduction

The global spend on alternative medicines is $60 billion a year and rising. In France, 75% of the population has used some form of what is often also called complementary medicine. That figure is around 50% in the UK, 42% in Canada, and 35% in Norway. More than three-quarters of German pain clinics offer acupuncture. Australians spend A$2.8 billion and the Europe-wide market for herbal remedies is over €600 million and growing, while Americans spend as much as $47 billion each year on what they know as alternative therapies. Complementary medicine, alternative medicine – call it what you will – unorthodox treatments are now the fastest growing sector of many health care systems.

Something is happening to the way we think about our health. Not quite a revolution, more of a sea change, a shift away from our being passive recipients of doctors' wisdom towards becoming active participants in our own health care. Patients arrive at the doctor's surgery armed with their own views on how their bodies work and what can be done to heal them.

Walk into a health food store in Dunedin, the southernmost city in the world, right now, and you can read the latest (winter sport 2004 as I write), edition of *Health and Herbal News*. It takes issue with the idea that prescription medicines are safe and duly approved by the New Zealand Government: 'regretfully the truth is far removed from perception'. Flicking through you might spot the article about the way 'most prescription drugs don't work', or the one about how you can 'ease stomach discomfort with slippery elm'. A thorough read reveals a section called 'Research Review' complete with reports of 'scientific studies (which) prove garlic's effectiveness'. Anyone across the Pacific leafing through the Manhattan Yellow Pages or those of Oakland, California, to the category Physicians and Surgeons, will find listed entries for Acupuncture, Alternative Medicine, Chiropractic, Holistic Health, Homeopathy, Naturopathic and

Osteopathic Physicians alongside those of Pediatrics, Hematology and Gynecology. On the other side of the Atlantic, a leaflet pushed through the letter boxes of Islington in north London advertises a newly opened suite of therapy rooms – 'a stunning holistic centre' offering an 'exceptional and diverse range of complementary therapies' including Craniosacral therapy, Energy healing, Metamorphic technique, Reflexology and Reiki.

This is probably to be expected in the rarefied districts of Islington or Manhattan, home to well-heeled baby-boomers. Dunedin, though, is a city of more modest means. Perhaps even more striking is 'Dr & Herbs', a small shop selling Chinese remedies and offering acupuncture in Bluewater, a new, and vast shopping mall south east of London. Bluewater expressly caters for a *mass* market. In the UK complementary medicine is now a key retail commodity. Boots – one of the best known drugstore chains, with operations in 130 countries – began selling herbal preparations and aromatherapy oils in 1991. In December 2002 one of the major UK supermarkets, Tesco, was reported to have bought a majority share in a prestigious London complementary medicine clinic (established 1987) that was, incidentally, opened by the Prince of Wales.

Alternative or complementary medicines and therapies have become a branch of health care. Driven by consumer demand, only marginally regulated and offering therapies that many scientists reject as absurd, these 'treatments' are mounting a challenge – not easily ignored – to several major aspects of medical care, from means of delivery to modes of action.

Many of the treatments bundled together under the heading of 'alternative' are far older than the conventional medicine they are supposed to complement. Acupuncture dates back thousands of years; likewise massage and reflexology. Homeopathy began at the end of the 18th century, long before antibiotics and heart transplants. Herbalism is perhaps the most ancient of all and certainly pre-dates the evolution of humans.

Our close relatives, chimpanzees and gorillas, eat several medicinal plants and seem to have an understanding of which diseases they alleviate.

Modern medicine is the new kid on the block, and a very successful one at that. For a while it looked like what we now call conventional medicine had swept away all before it, at least in the developed world. In just the last half century, antibiotics, vaccines and surgery have saved countless lives and transformed innumerable others. And yet complementary therapies are staging an unstoppable comeback.

Behind the scenes is a tussle. On one side are those proclaiming the virtues of complementary therapies; on the other, those deriding them as unproven, potentially harmful, nonsense. To complicate things further there is a comparatively recent addition to the fray: integrated or integrative medicine that attempts to merge the best of both worlds.

In the midst of all this are claims and counter-claims about what kinds of therapy do or do not work. One faction wants to place the body's own ability to heal itself centre stage. Another feels that therapies should be independent of state of mind. Another argument is between those who reject many complementary therapies on the basis that they are totally unscientific and those who argue that they might be using as yet unexplained mechanisms of action. Yet another is between those who want to put individualized care at the heart of medicine and those who believe that producing broadly applicable treatments is the way forward. There is even a debate around the question of what does 'work' mean with respect to any treatment?

This thrust and counter-thrust of ideas raises important questions itself. What are these claims based on? How specialized is the underlying thinking? What kinds of science are involved? What methods are being used to justify the claims? These questions are where this book starts, and working out some possible answers are what it is about.

The range of different ideas being brought to bear in the clash between alternative and mainstream medicine is remarkably wide, from immunology and neuroscience through clinical research techniques, pharmacology, sociology, anthropology and a good deal of epidemiology. The names of the sciences involved are comparatively unimportant. What is crucial, though, is the potential for understanding that each discipline and approach offers and the arguments over their relevance, strengths and limitations.

This book does not join in the tussle. Rather it stands on a hilltop overlooking the arena trying to see and report back on who is grappling with whom and how, and (tentatively) what might be making headway. This book is not going to answer the question 'does acupuncture work for back pain?'. It will, though, shed light on why we do not yet have any good answers to that question.

This is perhaps a more difficult approach, but I hope ultimately a more useful one. There is a saying: 'Give a man a fish and you will feed him for a day. Teach him to fish and you will feed him for life'. The plan is that you'll be better equipped to fish in the swirling waters of complementary medicine by the end of Chapter 10.

○

Complementary therapies have become sufficiently big business for there to be commercial clients interested in analyses of the market. In a report entitled 'Alternative Health care 2003', published by the KeyNote Ltd market research company, some figures highlight just how big the sector is. The market for herbal and homeopathic remedies and for aromatherapy oils only is 'believed to have grown 10% to

15% *per year* throughout much of the 1990s'. While it fell back in 2002 and probably 2003 due to a change in European regulation, the prediction is that it will rise 'to over 6.9% in 2006 and 6.5% by 2007'.

Look, too, at the growth in the number of complementary and alternative therapists practising around the developed world. New Zealand is typical: the New Zealand Charter of Health Practitioners, representing some 8,500 complementary/alternative practitioners, estimates that there are approximately 10,000 complementary practitioners in a country of fewer than 4 million inhabitants. On the other side of the world, there are more than 31,000 practitioners in the records of the European Committee for Homeopathy, while the UK's Shiatsu Society, formed in 1981 with just a handful of members, now has 1,730. There has been an explosion in the types of therapy available: massage, chiropractic, osteopathy, acupuncture, biofeedback, herbal remedies, homeopathy, radionics, naturopathy, reflexology, spiritual healing, water cures, cupping, iridology, hypnotherapy and more.

Official Australian government statistics reported at least 2.8 million Traditional Chinese Medicine (TCM) consultations (including acupuncture) per year in the country, with an annual turnover of A$84 million. More than 60% of Australians use at least one complementary health care product per year, including vitamin and mineral supplements as well as herbal products, and overall Australians spend about A$2.8 billion per year in the complementary sector – A$800 million on complementary medicines alone. Imports of Chinese herbal medicines to Australia have increased 100% per year since 1993.

The estimates of how many people use complementary medicines around the world vary – in part because data in each country are not collected in the same way and the definitions of complementary or alternative therapy are not consistent. Some sources have 75% of the French using some form of complementary or alternative remedy, whereas other sources say 50%; the percentages for the USA vary from 40% to around 70% and so on.

The same KeyNote report records that across other European countries the proportions taking complementary or alternative medicines vary from 50–60% in The Netherlands to a little less in Switzerland at 40%, with Belgium and Sweden quite close at 30% and 25% respectively; the UK trails with 20%. Various surveys and polls suggest that, broadly speaking, more women than men turn to these therapies. It also appears that the highly educated are of a more complementary bent. Whatever the size of the explosion, the trend is towards including new therapies, rather than replacing old ones. Few people are abandoning orthodox medicine; they are simply using complementary medicine as well.

At the same time, many doctors are embracing complementary medicine. In 2003 the Medical Care Research Unit of the University of Sheffield compiled a report for the UK Government's Department of Health showing that 49% of GPs – family practitioners – offered some sort of alternative treatment, with the majority offering it on site rather than referring to outside practitioners. Some doctors even see these therapies as a way of meeting their government-set targets. In Germany, which has a strong tradition of complementary and orthodox medicines running side by side, many doctors are also homeopaths. Numerous US family practices offer acupuncture, massage, aromatherapy and the like. Research quoted by the Australian Medical Association indicates that nearly half of the GPs included in a survey said they were interested in training in fields such as hypnosis and acupuncture, and over 80% had referred patients for some type of complementary therapy. There are now at least 29 academic journals on the topic and around 50 degree- or diploma-level courses in complementary therapy in the UK alone. In the USA at least 20 higher education institutions offer some form of complementary or integrative medicine courses; there are eight in Australia and upwards of 40 across Europe.

The media, too, have discovered complementary medicine. At one end of the spectrum is the sober, sceptical, view illustrated by a short piece in *The Washington Post* in spring 2004. It reported on an article in the *American Cancer Society Journal* highlighting a range of apparently useless alternative cancer cures and argued for better education for doctors and patients about such claims. In a similar vein, the UK's *The Times* has for over a year had a regular column called 'Junk Medicine', written by its science correspondent. A recent edition pointed out that the vast majority of alternative therapies have not been through the same strict clinical trials as is now required for prescription drugs, and of those that had, most failed to show any significant effect. A slightly different approach is offered by the *Guardian* newspaper, which features a regular column by Edzard Ernst, Professor of Complementary Medicine at the University of Exeter. A doctor by training, Ernst argues that alternative therapies should be carefully and rigorously tested. He is applying the conventions of medical science to what, to some people, are the more nebulous claims of the therapies, and finding some effective but many wanting.

At the other end is the human interest type of media coverage, which at times gives an impression of alternative therapies dealing in miracle cures. *The Times* Saturday Health Supplement, called 'Body and Soul', in which the 'Junk Medicine' column appears, most weeks also features a personal account from someone who had an intractable condition that conventional medicine was unable to treat and found relief only from some form of complementary medicine. I have to declare an interest here as I write a short piece that goes alongside these features examining what, if any, scientific evidence exists to support the treatment. I often have to report that there is simply not enough evidence to be able to draw anything but tentative conclusions.

The real media explosion in complementary coverage has been in magazines.

Gill Hudson, currently Editor of the BBC publication *Radio Times*, has been editing magazines, including *Fitness*, *Company*, *New Woman* and *Eve*, for more than 20 years. Hudson launched the men's lifestyle magazine *Maxim*, now the largest circulation publication of its type in the USA. She traces the rise in interest in complementary and alternative medicines back to the aerobics boom in the 1980s. Health and fitness became something that we could all aspire to and attain, says Hudson, rather than being the privilege of elite athletes. Publications sprang up to cater for this demand and the market began to grow.

Women's magazines started to change too. From their 19th century beginnings they had health pages, but these tended to be written by doctors and conveyed an air of authority, handing down wisdom from upon high. Some two decades ago, editors realized that readers wanted to get involved in their own health care and so started to provide tips for them to do so. Features on alternative therapies began to appear, and gradually treatments that had been considered counter-cultural or just quaint, such as herbalism, aromatherapy or shiatsu, moved to the fore. Today, says Hudson, alternative therapies are an essential element of all women's and lifestyle magazines. In fact, she doesn't quite see why they are called 'alternative' at all, so established are they in mainstream magazine publishing.

Hudson identifies the ageing of the baby boomer generation as one of the key drivers of this change. Now in middle age, this group were young in the 1960s, when authorities of all types were being questioned. While their parents would never have challenged a doctor, no matter what they were prescribed, baby boomers not only question them but go off and seek other advice if they are not satisfied with the answer. Furthermore, they have come to expect to live a good, long life and are not prepared to 'give up' when age starts to take hold.

And then there is the Internet. Even the most cursory web search turns up thousands upon thousands of alternative medicine sites, some clearly well researched and authoritative and some barmy by any criterion. There are also plenty of sites offering advice on conventional medicine, providing considerable detail of near enough the complete gamut of conditions, causes, prognoses and types of remedy. These are the modern equivalent of the sections on 'Diseases, Cure and Prevention of' (*The Home of Today,* published by the Daily Express) or the chapter entitled 'A Medical Dictionary' (*Newnes Everything Within: A Library of Information for the Home*) of popular domestic handbooks of the 1920s and 30s. Add this to the gradual reduction in deferential attitudes towards medicine that became noticeable during the 1970s and 80s and patients are often arriving in doctors' surgeries with lists of questions based on their Internet searches.

'Patients with cancer and other life-threatening conditions often turn to complementary/alternative medicine for a variety of reasons, and a major source of their information is the Internet', wrote cancer specialist Scott Matthews of the University of California in San Diego in the March–April 2003 edition of the journal *Psychosomatics*. In response, Matthews and his team have developed a series of questions to help patients determine the reliability of information on cancer information web sites. The answers to questions such as whether the treatments were for sale online, if the treatment was touted as a 'cancer cure' and if the treatment claimed to have 'no side effects', raise or lower metaphorical red flags – the more flags a web site has, the less reliable its information.

This is a noteworthy attempt to determine the scientific veracity of particular web sites. But what such a questionnaire cannot do is pass comment on the vast amount of patient testimony available. Virtually every complementary and alternative therapy web site, whether attempting to provide dispassionate information or to sell you something, will offer patient testimonials describing the effects of their particular treatment. These have a common theme, which goes something like this: 'I had a condition that was making my life a misery, and the doctors could do little for it. Then I discovered treatment X and I have never looked back'. They can be pretty compelling, particularly to someone suffering from a similar condition.

In her 1980s study of the coverage of medicine in the media, *Doctoring the Media: the Reporting of Health and Medicine* (London: Taylor & Francis, 1999), Anne Karpf noted that mass media treatment of alternative therapy was changing and was no longer as unsympathetic as it had been. Indeed, media support for complementary medicine could be seen as part of an attempt by editors to side with the 'voice of the people' against the domineering medical establishment. Three years later, Clive Seale's study of *Media and Health* (London: Sage, 2002) offers a rather different angle. He argues that the mass media counterbalances its reporting of health scare stories with 'the spectacle of ordinary people displaying exceptional powers when threatened by illness', a genre into which coverage of complementary and alternative treatments readily fits.

○

Complementary medicine has been described as the first patient-led form of health care. It is used most often for chronic health problems like lower back pain, eczema, stress or arthritis, which are not life-threatening but are conditions by which conventional medicine is regularly stumped. More recently, complementary health practices have increasingly been accepted and integrated into palliative care where the aim is not to cure but to comfort those with terminal disease. It seems that doctors in this field are more comfortable with a multidisciplinary approach.

As well as offering succour for intractable conditions, complementary therapies appeal to patients' dissatisfactions with orthodox medicine. In the book *Alternative medicine: Should we swallow it?*, Tiffany Jenkins and her colleagues list a number of the reasons why this might be. They cite disillusionment with: being treated as machines needing to be 'fixed'; reliance on 'artificial' pharmaceuticals with unacceptable side-effects; and short consultations that process people like a factory conveyor belt. By contrast, a session with a complementary therapist will usually last around an hour. The philosophy of these therapies is to empower the patient, giving them an active role in identifying problems and solutions. Most particularly, complementary and alternative medical approaches are typically holistic – concerned to treat the whole person, not simply the specific symptoms of components needing repair.

○

A further explanation about the use of the words *complementary* and *alternative* is necessary. Until about 10 years ago most people giving or receiving the therapies discussed here would have called them 'alternative'. But there has been a move to describe them instead as 'complementary' to emphasize that they are intended to run alongside, rather than in opposition to, orthodox medicine. This description is more common in the UK, Europe and the antipodes than in the USA where 'alternative' therapy is still the most widespread term. Doctors, researchers and many practitioners have overcome this confusion by referring to 'complementary and alternative medicines', abbreviated to 'CAMS'.

The definitions are not clear-cut. The House of Lords Select Committee on Science and Technology's Sixth Report on Complementary and Alternative Medicine reads 'Complementary and Alternative Medicine (CAM) is a title used to refer to a diverse group of health-related therapies and disciplines which are not considered to be a part of mainstream medical care'. The inquiry on which their report is based was set up in the wake of recognition by the UK government that the use of complementary medicine was growing both in the UK and elsewhere across the developed world.

An article in Melbourne's *The Age* newspaper in March 2004 discussed the increase in interest in complementary and alternative medicine in Australia as follows: 'Most doctors would agree that alternative medicine should be approached cautiously. But there is less consensus about "complementary medicine", which the Australian Medical Association describes as embracing acupuncture, chiropractic, osteopathy, naturopathy and meditation – or even less mainstream treatments such as aromatherapy, reflexology, crystal therapy and iridology – used in conjunction with conventional medical treatment'. Here there appears to be a sharper distinction between complementary and alternative than in the House of Lords Report.

Commercial organizations have different agendas and so yet other definitions. The KeyNote report is designed to help investors and businesses, and thus excludes what it describes as 'recreational pursuits', such as yoga and Feng Shui; some types of massage; and systems or disciplines with a religious or spiritual aspect, such as faith healing.

What we have today is a picture – which will no doubt continue to change – wherein the kind of medicine called

'Western', 'scientific', 'conventional', 'orthodox' or 'allopathic' enjoys a distinct advantage, a sort of top-dog status compared with others which are marginalized as 'different', 'unorthodox', 'alternative' and so on. This picture has evolved over a long time. In the UK, for instance, the medical *profession* could be said to have begun with the Medical Act of 1858, which specified what qualifications allowed people to describe themselves as doctors. Critically, the Act distinguished between qualified and unqualified practitioners, but did not stop the latter from practising. A statutory boundary was created to be policed by the General Medical Council, the body set up by the Act. The story was different in detail and dates in the USA, Australia and elsewhere, although it is about the same issues: licensing and (above all) control of who can and who cannot call themselves a doctor.

Even depending on this definition – treatments employed by registered doctors are orthodox and others are complementary – brings problems. For instance, chiropractic is regulated by law in the UK and homeopathy is deeply integrated into the orthodox medical profession in Germany, so do they count as complementary or orthodox therapies?

The late Roy Porter, historian of medicine *par excellence*, took this view: 'In a medical world which is increasingly bureaucratic and technology-driven, the Hippocratic personal touch seems in danger of being lost'. Confidence in the medical profession had been undermined, he posited, driving the renaissance, since the 1960s, of 'irregular medicine', a term some two centuries old.

> The eighteenth century was arguably the golden age of 'quackery' – a loaded term, for when speaking of non-orthodox medicine we should not automatically impugn the motives of the irregulars nor deny their healing gifts. Far from being cynical swindlers, many were fanatics about

> their techniques or nostrums.... From the 1780s the one medicine which would truly relieve gout – it contained colchicum – was a secret remedy: the *Eau médicinale*, marketed by a French army officer, Nicolas Husson, and derided by the medical profession.
>
> (Roy Porter, *Blood and Guts: a short history of medicine*. London: Allen Lane, 2002).

As Porter illustrates, therapies can move from being classed as alternative to orthodox over time and back again.

I have used *complementary* and *alternative* more or less interchangeably throughout this book. Nothing is implied by referring to one technique as alternative and another as complementary – I am merely acknowledging that each is not recognized as part of the pantheon of orthodox medicine. Most of the references in the bibliography refer to CAMS, but I have chosen to use as few acronyms as possible – I don't like reading strings of letters and have no wish to impose them on anyone else!

○

The list of therapies that come under the broad heading of complementary and alternative is large and growing. Likewise, there are a number of ways of classifying this wealth of treatments. They can be divided into physical techniques such as osteopathy or massage; qi (or chi) energy-based such as shiatsu or reflexology; mind-based, such as hypnotherapy or neuro-linguistic programming; or even geomancy, such as crystal healing. The therapies can also be categorized by their origins. Acupuncture and shiatsu are based on the Traditional Chinese

Medicine concept of energy meridians running through the body; psychotherapy has emerged from the western tradition of Freud and Jung. None of these categories is particularly satisfying, as there is frequent crossover of ideas from one to another. This is not unique to alternative therapies: biomedical disciplines are equally fluid and ideas cross from one to another all the time. Genetic factors in heart disease have been uncovered by epidemiology and finessed by geneticists, while cardiologists and general practitioners use the information to treat patients.

The therapies discussed here are used as illustrations. When demonstrating the problems involved in evaluating therapies there is little point talking about those for which very little data exists. Therefore, all the complementary or alternative treatments mentioned in this book have one or both of the following features: they have been subjected to some form of research into their effectiveness or they are being used in significant numbers alongside orthodox doctors in orthodox medical practices. These include chiropractic, osteopathy, acupuncture, homeopathy, Bowen technique, acupuncture, psychotherapy, shiatsu and reflexology. This is a small list compared to the huge, and growing, number available.

Furthermore, this relatively short list implies no judgement either way about the effectiveness of other treatments. The problems of assessing and measuring are just as relevant to aura balancing or bioenergetic stress testing as they are to acupuncture. The absence of a therapy from this book merely reflects the fact that there has been far less, if any, research into that therapy, or that it is rarely included in integrated medicine.

A quick word about how research is done is needed at this point. For the results of a study to be acceptable they have to be published according to a quality control procedure known as peer review. This simply means that before an editor will accept a paper for publication it has to be refereed by other academics with similar expertise. The impact of that research is, in part, dictated by the journal in which it appears. There is an acknowledged pecking order of journals, with the elite typified by ones such as the *Journal of the American Medical Association*, the *British Medical Journal*, the *New England Journal of Medicine* and *The Lancet*. Research featured in these publications is hard to ignore; the corollary of which is that research published in journals further down the pecking order is correspondingly easier to ignore. That said, all the articles quoted in this book are from peer-reviewed journals.

There is one major omission from the list of therapies discussed here: herbal remedies. These are biologically active medicines. They can be tested in more or less the same way as pharmaceutical drugs and their efficacy is as easy, or difficult, to determine. The debate surrounding herbal medicines concerns regulation, safety and conflicts with other prescription drugs that patients might be taking. A significant number of pharmaceuticals available today have their origins in plants, which biomedicine has well established ways of exploiting. While herbal medicines may be classified as alternative or complementary, they are similar for the purposes of testing. It is how to test *dissimilar* therapies that is the theme of this book.

○

Discoveries are made at the limits of scientists' abilities. Physicists push their giant particle accelerators to ever higher energies; biologists delve deeper into the workings of our cells; and astronomers stretch the range of their telescopes to see further across the vastness of space. Complementary and alternative medicines are difficult to study, they require a reach into unknown territory. Like all thriving areas of investigation there are factions, personal animosities and a great deal of passion. There are believers and sceptics, waverers and staunch defenders, advocates and rejectionists – never mind the indifferent. This book is the story of how the latest research into complementary medicine, practitioners and patients is giving medicine itself a thorough examination.

2

medicine's conundrum

I received a package through the post yesterday, a new gadget for my collection of supposedly useful technology. As always I got out my penknife to cut the tape sealing the box. I was tired and nicked my thumb while closing the knife. This was irritating, but hardly life threatening; the plaster I put on it was more to prevent the blood going everywhere than to help the cut heal. True, it might have become infected but that was unlikely and not a real concern.

Had I cut myself 100 years ago I might not have been quite so relaxed. My body's ability to mend itself would have been the same then as it is today, but there was one crucial difference. Were a wound to become infected in 1904, little could be done if my own immune system failed to fight back. A minor cut could kill if it became infected, and often did. Fortunately the past century has seen one of the most significant medical breakthroughs: the discovery and development of antibiotics.

Antibiotics have reduced the threat from infectious disease dramatically. Diseases that used to devastate populations, such as cholera, typhoid and even plague can be tackled if there are enough antibiotics to go around.

The other huge advance is vaccination. The first vaccination, for smallpox, is credited to Edward Jenner in 1796. It was a hundred years before the next one, against rabies, was developed, followed over the next 50 years by vaccines for plague, diphtheria, whooping cough, tuberculosis, tetanus and yellow fever. With a few major exceptions, what antibiotics can't kill, vaccinations can prevent. Virtually everyone in the developed world receives a series of immunizations as a child that saves them from a whole host of potential killers – including measles, polio, tuberculosis, whooping cough and diphtheria. The feather in the vaccinators' cap is the eradication of smallpox. With few pocked faces any more in the West, there are barely even reminders of a disease that used to kill 30% of the people it infected.

Polio is next on the list. The World Health Organization hopes to eradicate it within the next few years. The success of the polio vaccination campaign is most visible, or more accurately, invisible, in the West. People of my father's generation lost schoolfriends to iron lungs after they had picked up the polio virus: it attacks the muscles and can leave victims unable to breathe unaided. John Prestwich is 65 – retirement age in the UK where he lives – and holds the record for the person who has lived longest in an iron lung. Advances in technology have provided him with a portable device rather than an enclosed canister. Hospital wards full of rows and rows of iron lungs have gone forever, and the last few machines are kept for emergencies only. Gone too are the leg braces, limps, wheelchairs and withered limbs that were the most visible reminders of the muscle-destroying infection. Thanks to vaccination, the disease has been wiped out in the developed world. While not yet the end of polio – pockets persist in South Asia and Central and West Africa – this is a significant marker on the way to its eradication.

Until the 1960s it was assumed that scientific advances were largely responsible for the increase in lifespan. Then Thomas McKeown, Professor of Social Medicine at Birmingham University in the UK, suggested that it was improvements in public health and nutrition that had had the bigger impact. McKeown argued that the provision of clean water and proper sewage disposal, the destruction of insanitary slums and the availability of a better diet were responsible for people living longer. There is a very close correlation between the availability of clean water and better sanitation and the reduction in the incidence of water-borne diseases such as typhoid and cholera. Better living conditions have drastically reduced the incidence of diseases like tick-borne typhus, which thrive in crowded housing.

It was a bold claim and appeared to relegate biological science to a bit part in the theatre of human health. Today, though, the

McKeown thesis is seen as missing some important elements, not least because he did not consider the role that doctors might have had in helping people to improve their diet and hygiene. McKeown also ignored the contribution that doctors have made to public health by working to enhance living conditions – sitting on public committees, lobbying politicians and so on.

The debate rumbles on regarding exactly which elements have made the biggest contribution to increasing lifespan: medical interventions, public health or nutrition. Science, though, has contributed to all of these. Biologists lead the way in determining the nature of infectious diseases, discovering, for example, that cholera is the result of infection with a waterborne bacterium and malaria is the result of a parasite passed on by the bite of a mosquito. Without that knowledge these diseases would be far harder to tackle. Malaria, for example, was so named as it was originally thought to be the consequence of breathing bad – 'mal' – air. Closing the windows at night to keep out 'bad air' would have had some success at stopping mosquitoes biting. But it wasn't until British Army doctor Ronald Ross discovered that mosquitoes transmitted malaria that the disease could be fought by preventing the insects from breeding and biting.

Another development that has saved lives is the improvements in emergency medicine. Individuals can now recover from previously fatal traumas. Procedures vary from sophisticated surgery that can reattach damaged limbs to the simple use of pressure on a wound to stop bleeding. This has replaced the old idea of a tourniquet, which was shown to increase the chance of gangrene, which in turn could kill. Even basic first aid training now includes resuscitation techniques that can help people survive a heart attack.

And then there's transplantation. The idea has been around for centuries – replacing worn-out bits of our bodies with parts from another human. Or even from another animal – called xenotransplantation. Xenotransplantation was first tried in the 17th century when bone from a dog was used to repair the injured skull of a Russian nobleman. It is now undergoing a controversial resurgence due to our ability to genetically engineer animals. The hope is that pigs or other animals can be genetically engineered so that their organs, or cells even, resemble human ones, boosting the supply of donor tissues. That is still some way off, and today the only animal organs widely used in human transplantation are pig heart valves.

The number of different organs that can be replaced in humans is extraordinary. Lungs, hearts, kidneys, corneas, livers, pancreases, skin, bone marrow and even entire bones are harvested from living or dead donors and swapped for diseased organs, to extend and improve the recipients' lives. There have even been two attempts to transplant arms and hands, with limited success, and a few doctors are seriously considering transplanting entire faces. We are discovering that human organs can be treated pretty much like car components. You can replace worn-out parts as long as you ensure that the new ones match.

The ingenuity of the surgeons is coupled with that of drug developers. One drug in particular, called cyclosporin, is responsible for more successful transplants then any other. It is an immune suppressant: it tones down the body's natural defences, preventing them from attacking the transplanted organ.

Like an army, the human immune system has a reconnaissance arm that scours the body for invaders. On finding one, it calls in the big guns to destroy the intruder. This is what can happen to a transplanted organ: if the immune system recognizes it as foreign, it will be attacked and killed – rejected, in other words. To prevent this, organs are matched as closely as possible to the recipient; the development of sophisticated matching techniques has greatly improved the success of transplant operations. But a perfect match is possible only between identical twins, so there is always a chance that a recipient will reject a new organ. This is where cyclosporin comes in. It dulls the immune system's senses, allowing a well-matched organ to thrive.

The future of transplantation is even more extraordinary. Alongside xenotransplantation research, the technology is being developed to build organs from our own cells and so avoid rejection. Laboratories across the world are trying to grow artificial organs using many different techniques. Some are persuading cells to take up residence in delicate scaffolds of natural materials such as coral, or are creating artificial ones out of synthetic materials. Others are finding ways to harness our cells' ability to organize themselves into complex organs and tissues. This discipline, called tissue engineering, could provide a way to repair damaged nerves or muscles or even to grow entire new kidneys.

Any whistle-stop tour can only hint at the breadth and sophistication of modern medicine. Diseases that once were fatal have diminished or disappeared; people recover from horrendous accidents; and worn-out bits of the body can be replaced. The major advances in medical science have changed societies and expectations. Birth rates in the developed world have plummeted as it has become the norm for children to survive to adulthood. There are few children left brain-damaged because of measles and few families devastated by the death of half their children from diseases such as cholera and typhoid.

Our confidence in the power of modern medical science to provide a cure is a testament to its success, but it has also left it with a problem. It is powerful, but it is not omnipotent. Many diseases still defeat the ingenuity of every physician, specialist or surgeon. Large and increasing numbers of people are suffering from conditions that are either difficult to treat or incurable. These diseases are principally ones of either prosperity or maturity, so they could only appear in significant numbers once the infectious killers had been wiped out, at least in the developed world.

The World Health Organization's database of death rates and causes stretches back to 1950. Even a quick glance reveals the radical change that has happened over the last half century. The average lifespan has increased by 25 years, and that has brought a change in the illnesses that afflict us. As deaths from infectious diseases have plummeted, those from chronic conditions such as heart disease and cancer have soared. In the USA in 1900 the leading causes of death were pneumonia, tuberculosis, diarrhoea and enteritis, accounting for around 35% of all deaths. By 1999 heart disease killed 32% of the population and cancer 24%, with stroke in third place on 7%. These figures are mirrored across the developed world.

There are over 300 different types of cancer, all with very different symptoms, treatments and prognoses. People with cancer of the breast, uterus or testis have a 75% chance of surviving more than five years, whereas those with liver, stomach or lung cancer have less than a 15% chance of surviving that long. In fact, there are as many different types of cancer as there are types of cells in our bodies. The reason is simple: cancer is the unregulated growth of a single cell. If the out-of-control cell is from the lung, the result is lung cancer; if it's a colon cell then colon cancer ensues. And just as lung cells are very different from colon cells, so lung cancer is very different from colon cancer. Worse still, there are almost as many types of lung and colon cancers as there are types of lung and colon cells.

The wide range of different cancers is just the start of the problem for scientists and doctors. Infectious diseases are caused by invasions, be they of bacteria, viruses or parasites, such as bacterial meningitis, influenza or tapeworm, respectively. Most of these organisms have a physiology that is different from ours, giving drugs something to aim at. Penicillin attacks certain types of bacteria, but fortunately does nothing to human metabolism – apart from those who are allergic to it.

The problem with cancer cells is that by and large they are very similar to normal cells; drugs that will affect them will also harm healthy tissues. This is why many anti-cancer drugs can have such devastating side-effects. It is extremely difficult to find something that will hit just the cancer and leave the rest of the body alone.

The most common types of anti-cancer drug home in on one of the cancer cells' few obvious differences – their faster-than-normal ability to multiply. A drug that targets dividing (proliferating) cells can weaken or wipe out a tumour. But side-effects are manifold. One of the fastest-dividing groups of cells is hair cells. Drugs aimed at rapidly growing cancers also hit these hard, which is why many people go bald during chemotherapy.

Another approach is to identify a weakness key to a particular type of cancer. Prostate cancer, for example, requires testosterone to grow, so drugs that block testosterone release can slow prostate cancer growth. Again there are considerable side effects: this hormone treatment is chemical castration. Men receiving it lose their body hair and can develop breasts and suffer menopause-like hot flushes. And in the end the treatment fails: the cancer becomes insensitive to the hormone and grows regardless, often with fatal results.

Cancer is a local uprising, not an invasion. As any general will confirm, it is far harder to quash insurgents than to fight an alien attack. Autoimmune diseases, another type of 'cellular uprising', pose similar problems. There are forty or fifty different autoimmune diseases, most of which appear to be on the increase and the prevalence of which varies from population to population. An interesting indication of their growth is a document published by Theta Reports, an independent publisher of market reports for the health care sector. Called 'Autoimmune Disease Therapeutics Worldwide', it predicts that the global market for autoimmune disorder treatments is growing by 15% per annum and is expected to reach over $21 billion by the year 2006.

In rheumatoid arthritis the immune system slowly eats away the delicate linings of the joints, resulting in an agonizing grinding of bone on bone. Joints swell, movement becomes painful and patients become increasingly immobile. To stretch the military analogy even further, it is like a rogue battalion attacking an innocent group of civilians. Why this happens is still a mystery.

Painkillers and anti-inflammatory drugs can help. When joints deteriorate too much, a range of artificial ones are available. But there is no cure. Of course, a great deal of research is being done on rheumatoid arthritis and the like, some of which will undoubtedly produce improved treatments with time.

There are many such autoimmune diseases that, like rheumatoid arthritis, have no cure: early onset diabetes, multiple sclerosis, bullous pemphigoid, psoriasis, ulcerative colitis, Grave's disease, pernicious anaemia and more. All are the result of our immune system turning on us and all pose similar problems: preventing disease without hampering our vital protective forces.

Heart diseases are yet another health care headache. While there is a strong genetic component to all forms of heart disease, much of the problem is lifestyle-based. The rise in availability of cheap, fatty, sugary fast foods combined with our increasingly sedentary habits is making obesity the biggest threat to human health of the 21st century.

Drug companies meanwhile are working hard to develop pharmaceuticals that will enable us to eat our fill and then pop a pill to stay slim and healthy. Some are already on the market, such as Xenical®, which prevents fat from being absorbed from the gut, or Meridia®, which works on the brain to suppress hunger. Others are in development, such as Sanofi-Synthélabo's rimonabant or GSK's compound 181771. It costs around one billion US dollars to bring a drug to market, so the investment represented in developing a weight loss pill is enormous. Clearly the financial brains within the pharmaceutical companies believe that it is worth spending the money. In other words, they are gambling that people are going to continue to be obese and that sufficient numbers will turn to chemical rather than lifestyle solutions to the problem.

The real fix to the obesity epidemic is much simpler: eat less, eat better, exercise more. It's a prescription that doctors are handing out worldwide and which most patients seem to be ignoring. The message is coming from everywhere. I act as consultant to a TV programme in the UK that attempts to use shock tactics to drive that message home. Almost everyone knows they need to look after themselves, but it's clear that many people struggle to make the change. I find it hard and I'm lucky: I enjoy vegetables, can drag myself to the gym, can afford decent quality food and know how to cook it. One of the biggest problems is that unhealthy, highly processed fast food is cheap and provides more calories per penny than fresh food. There is no doubt that reducing obesity will cut the numbers of heart attacks and strokes, and there is little doubt that simply telling people what to do is not working.

One of the greatest scourges of a longer-lived population is Alzheimer's disease, the cruel dementia that robs one in 20 people over 60 of their memories, movements and ultimately lives.

Again we know a lot about the molecular, behavioural and physical changes that occur during the progression of Alzheimer's. Yet, at present, there is little doctors can do to halt its progress. As another of the ailments where our bodies turn against us, it is proving extremely hard to treat.

There are more diseases to add to this list, particularly of children, that seem to be rising relentlessly in the affluent West. Take asthma. Statistics vary from country to country, but the published research agrees that the number of asthmatics has at least doubled in the last 20 years. A parallel rise in allergies has caused such concern that the European Union has funded a €29 million investigation, called the Global Allergy and Asthma European Network. Then there is Attention Deficit Hyperactivity Disorder, a behavioural disorder in children that has been linked to diet and allergy. It, too, is being diagnosed with greater frequency. Add irritable bowel syndrome, chronic fatigue and chronic pain, and the range of conditions for which, at present, medicine has little to offer begins to look rather long.

This, then, is medicine's conundrum. AIDS, TB and SARS notwithstanding, medicine has dealt with infectious disease in the West. Almost overnight antibiotics made lethal infections minor irritations and vaccinations took up the slack. These, though, were the easy targets. The diseases that afflict us now are the difficult ones. Chronic conditions, often born from failures in our bodies' own defences, are extremely tough to treat.

Its unlikely that Alzheimer's, rheumatoid arthritis, cancer and the like will ever have a quick, simple, cure analogous to a course of antibiotics. Chronic conditions tend to require long term treatments. Anyone prescribed statins to reduce their cholesterol levels will probably take them for life. Transplant patients have to have regular checkups and many take immune suppressant drugs for ever.

Yet to an extent we have been spoiled by a century of medical success. We expect a ready remedy for heart disease or a pill that will stop the dementia that is eroding our loved one's personality. Meanwhile, how many times has a cure for cancer been trumpeted in the media, and how many of our friends and relatives still die from its many forms? Medicine's success has backed it into an expectation-management corner.

What's more, one of modern science's greatest triumphs is highlighting just how hard it will be to tackle these chronic diseases. The Human Genome Project has identified all the genes that direct how a human develops. It has listed all three billion letters of code that make up our genome and identified around thirty thousand active genes. Many geneticists believe that reading the genome was the easy bit. Understanding it is going to be a much bigger job.

One of the central findings of the Human Genome Project is that there are around one and a half million points in the genome that can differ from person to person. These are called single nucleotide polymorphisms. Everyone has the same number of genes; it's these individual letter changes that make the difference, just as swapping the letter 'i' in 'drink' for a 'u' gives you 'drunk' – a very similar word with a very different meaning.

One corollary of understanding our differences is 'individualized medicine', currently the focus of heavy investment. We know that some people benefit from certain drugs while others find them useless, or even harmful. For example, 30 to 50% of patients with clinical depression do not respond to the drugs prescribed, increasing their chances of dying from a depression-related condition. One reason could be several genetic mutations which affect an enzyme in the liver called CYP2D6 involved in the breakdown of these drugs. A handle on genetic differences means we can start to identify which compounds fit which genetic profiles and so develop drugs tailored to individual needs.

No genetically tailored drugs are on the market as yet, but it is only a matter of time. Iressa®, from pharmaceuticals giant AstraZeneca, is one of the few treatments available for small cell lung cancer. Unfortunately it only works for some patients. Research published in the spring of 2004 by two independent groups has pinpointed a genetic difference between patients that respond well to Iressa and those that do not. It should now be possible to develop a genetic test to determine whether or not the drug should be used for a particular patient. That may not be hugely significant on its own – it only takes two weeks to find out whether Iressa works anyway. The implications, though, are far wider.

Discovering that possession of a particular form of a gene dictates whether a drug will be effective allows researchers to home in on the weaknesses of small cell lung cancer. This opens up the possibility of developing other treatments that can either bypass the genetic element or enhance it. Either way, this extra knowledge, gained from genetic research, should help researchers develop more effective treatments for one of the leading causes of death worldwide.

The impact of genetics will also be felt in chronic conditions. An example has already emerged. The group of drugs known as statins are used to control levels of cholesterol in the bloodstream. High blood cholesterol levels have been linked to chronic heart disease and patients prescribed statins usually have to take them for the rest of their lives. A study published in June 2004 in the *Journal of the American Medical Association* showed that the drugs will be significantly less effective for those with two specific genetic mutations. The authors of the paper write, 'We recognize that these data have considerable pathophysiological interest and provide strong clinical evidence that there may be promise in the concept of "personalized medicine"'.

One potential obstacle to all of this is that the business of drug production is the economics of scale writ large. Pharmaceutical companies make back their billion dollar new-drug spend by marketing it to 10% or more of the world's population. Tailored drugs sound great, but how are they going to be profitable? Are manufacturers going to make up 20,000 versions and ship supplies around the world? Are street corner chemists supposed to have the expertise to mix up the right combinations, or will each doctor have a machine on their desktop? As yet there are no clear answers to these questions.

○

The march of medical science has greatly improved the chances of surviving previously fatal conditions, made hitherto debilitating ones easier to live with and raised our expectations of health care. More and more is being asked of doctors, and one area that has attracted consistent criticism, especially in public sector funded health systems, is the time that doctors can spend with their patients. Figures recorded in 2002 for the average length in minutes of a consultation with a family or general practitioner were: Germany 7.6; Spain 7.8; UK 9.4; Netherlands 10.2; Belgium 15.0; Switzerland 15.6. Figures recorded in 1998 for the USA gave an average of 18.5 minutes. These snatched consultations are deeply dissatisfying for both parties. It is proving difficult to balance the development of hi-tech medicine with the basic, low-tech need of doctors and patients to spend time together.

The situation can be somewhat different with private medical care. Here the doctor is the direct employee of the patient and, broadly speaking, the time that doctor and patient have together is related to the patient's ability to pay. It costs a great deal of money and is not a privilege available to the majority. Medical insurance schemes pay for private health care, but control their spending and can and do impose limits on the amount of access a patient gets to a doctor.

Given the problems that orthodox medicine is encountering with treating the growing number of chronic degenerative diseases it is perhaps not surprising that we are turning in ever greater numbers to alternative therapies. One of the driving forces is the natural instinct to seek a cure. We have come to expect an effective treatment for illness and disease, and if one form of medicine, orthodox, cannot provide it then we will look elsewhere. The other incentive is that complementary therapies offer something more subtle than allopathic medicine. Medics are rarely able to spend the time to help a patient come to terms with living with a chronic condition. Long-term illness can put a great strain on limited resources and there is no 'cured' box to tick at the end – only, to be brutal, a funeral. Modern medical practice does not ignore chronic conditions (far from it), but they are just not its strong suit. Complementary medicine, on the other hand, thrives on prolonged involvement with patients and positively encourages them to return. The financial implications of this should not be ignored. Regularly returning patients provide practitioners with a reliable income and could, in theory, be exploited. It might, though, be considered similar to the old Chinese practice of paying your doctor to keep you healthy rather than restricting your visits to a few expensive ones when you are sick.

Complementary practitioners talk about 'wellness' and 'holistic' treatments, not always offering cures but instead suggesting ways of living with one's lot. They give advice on diet, exercise, pampering, and just generally being nice to yourself. A therapy that engages you and makes you feel part of the process of maintaining your own health is alluring, particularly when the alternative is to take a bunch of pills from a doctor whose whole demeanour says 'I'm stumped'.

It might be natural to assume that allopathic medicine has a lot to learn from complementary and alternative methods where it comes to chronic disease. That complementary medicine welcomes long-term treatments suggests that it is far better at treating difficult and low-level conditions. But many of the assumptions about this type of treatment are just that – assumptions – without any evidence to back them up. There is a wealth – no, an unimaginably large fortune – of anecdotal evidence that complementary approaches help people with chronic diseases to feel better. So too does long-term intensive nursing, which is firmly within the conventional medicine fold. Is there any added benefit offered by complementary and alternative therapies or are they equivalent to nursing care? This question is crucial and extremely difficult to answer. In fact there are probably a whole series of answers, but if any complementary therapies prove effective then allopathic medicine has something to learn; if not, then a billion dollar industry is based on nothing but good old-fashioned tender loving care.

3

the advocates

The most powerful advocates of complementary and alternative medicines are the millions of people who pay for them each year. The figures are extraordinary. Half the population of the UK has visited an alternative practitioner; so have half of Americans, more than half of Australians and three-quarters of the French. Around 3,000 French doctors, 5,000 Polish doctors and 7,000 German doctors are trained homeopaths according to the European Committee for Homeopathy, a Europe-wide association for homeopathy professionals. Users of alternative treatments might pay for alternative treatments once, but unless some need were actually being satisfied it is hard to see why they would continue to spend the vast amounts necessary to explain these figures.

The advocacy of two other groups is crucial to the mushrooming of complementary medicine: the practitioners themselves and the growing number of mainstream doctors who work alongside them. It is perhaps this second group, conventionally trained medical doctors, that is the most interesting. Many of the ideas contained within complementary therapies clash with modern science – ideas such as energy channels in the body, or the ability of an ultra-dilute solution to effect a cure. Yet many doctors, in seeming contradiction to their training, either practice some form of complementary treatment themselves or refer their patients to other therapists.

Doctors tend to be healers first and scientists second. The Hippocratic oath is largely about curing the sick, not understanding why they are ill in the first place. It recognizes science but puts a strong emphasis on humanity and caring. Consider this from a modern version of the oath written in 1964 by Louis Lasagna, Academic Dean of the School of Medicine at Tufts University: 'I will remember that there is art to medicine as well as science, and that warmth, sympathy, and understanding may outweigh the surgeon's knife or the chemist's drug'. It is this healing aspect of medicine that doctors sign up to first and foremost. There are of course those who lean towards research, and for them the desire to understand the diseases they treat is compelling, but for most of the doctors I've talked to, at least, understanding takes second place to helping their patients feel better.

The stereotypical complementary therapist is a rather vague character, possibly long-haired, trailing a whiff of patchouli and having a tendency to tie-dye. This is often far from the truth. Take Dr Mike Cummings, currently Medical Director of the British Medical Acupuncture Society. Mike is straightforward, clear, unsentimental and from a highly orthodox background. He discovered acupuncture in a rather unusual way while part of an institution that is as far from romantic hippiedom as it is possible to get.

As a medical student Cummings was particularly interested in musculo-skeletal conditions – sports injuries, strains, sprains, cramps and the like. Disappointed with the lack of emphasis on these types of common problem, he took himself off on sports medicine courses and spent time in rehabilitation centres. On qualifying he realized that it was not going to be easy to pursue his interests if he followed a conventional medical career. So he joined the Royal Air Force. Here was an organization full of fit, active individuals that gave him ample opportunities to develop his skills in sports medicine. He found plenty of injuries to treat and was enjoying learning how to do so. Then came a surprise. In Her Majesty's armed forces, Mike Cummings encountered acupuncture.

One Wednesday afternoon he was in the medical quarters on camp when he noticed that the door to the senior doctor's office was shut. This was unusual, so he asked the duty sergeant what was going on. 'The Squadron Leader is doing some acupuncture', came the reply. Cummings was stunned. What's more, the sergeant informed him, 'the Queen was paying for it'. The RAF, it turns out, has funds to allow doctors to train in a variety of different practices, and acupuncture is one of them. Armed forces around the world are known for their conservatism rather than their willingness to embrace unorthodox ideas. No wonder Mike was stunned.

Cummings did nothing with this information for a year or so but then found himself running a station medical centre and, his curiosity unsated, decided to take a course in acupuncture. He chose a short, practically based offering run by the British Medical Acupuncture Society; this gave him a grounding in how needles can be used to heal. The conventional medical approach to muscular injuries involves a lot of injections of drugs such as cortisone, and Cummings had used this more conventional form of needle insertion as a mainstay of his work, even though, as he admits, he did not always fully understand how it worked. Until that point, however, he had not considered that the needles themselves might be part of the effect. Trained in pharmacology, he assumed that it was the contents of his injections alone that did the work.

On returning to his practice Mike Cummings began using needles extensively. He found that they could be very useful for diagnosis, an extension of his fingers as he probed painful regions and mapped out the extent of the muscular damage. They also became part of his toolbox for treating the conditions he saw. Tense, bunched muscles could be released by inserting a slim acupuncture needle – an instant relief that he could see working before his eyes. Acupuncture became an important addition to his repertoire of treatments. His evidence was that he could relieve patients' pain quickly and easily. His own observations convinced him that there was something to the ancient idea of sticking needles into people to make them better.

When the time came to leave the Air Force he was unsure how to continue as a musculo-skeletal doctor, as that was how he still thought of himself, rather than as an acupuncturist. Then an unexpected opportunity arose to take over an acupuncture clinic. Naively he expected to turn the practice into a musculo-skeletal clinic and develop a portfolio career in medicine. Instead, the demand for acupuncture was so great that the clinic turned him into an acupuncturist. He now says that had anyone told him when he was in the Royal Air Force that he would become a complementary therapist he would have laughed out loud. Yet this upstanding military doctor has entered the world of complementary and alternative medicine.

Cummings then began to take an interest in the research that had been done on the efficacy and effectiveness of acupuncture. He read the papers, examined the studies and was amazed to find that, according to them, the treatment he used all the time did not work. The published findings were totally at odds with the results he saw in daily practice. Today, after many years involved in research, he reckons that the design of clinical trials is often at fault.

The evidence that convinces the pragmatic Cummings is not an abstract trial where patients are averages and numbers but his own eyes: his patients get better. This personal testimony is a common theme that runs through virtually all that advocates of whatever form of complementary or alternative medicine have to say on the matter, be they doctors, practitioners or patients. First person accounts of success are compelling.

Mike Cummings is one of many thousands of medical doctors adding a complementary therapy to their repertoire. The British Medical Acupuncture Society is a body for doctors and dentists, as well as other health professionals, who also practice acupuncture. But there are physicians who are chiropractors, osteopaths, reflexologists, shiatsu masseurs as well as all those who are homeopaths. In the UK six homeopathic hospitals operate with public funds and medical doctor–homeopaths are common throughout Europe. It is not necessary, though, to train in a particular therapy to encounter it. Many other doctors come into contact with alternative therapies daily.

Kate Kuhn is a General Practitioner, a family physician, based in Buckinghamshire near London. She has a PhD in biochemistry, reads the medical journals and does her best to keep up with the latest thinking. She is also very interested in complementary medicine and takes a broadly pragmatic view. The therapies may not work through any mechanism her scientific background has trained her for, but what benefits her patients is her primary objective.

Like virtually all those I have interviewed, Kuhn believes that complementary therapies have more to offer people with chronic conditions than those in acute emergencies. If a patient has a strangulated hernia or a severed limb, the best response is surgery. However, if someone comes to her with arthritis, for which orthodox medicine can offer very little, she is open to anything that might improve their quality of life.

An increasingly common issue for Kuhn, as with many doctors, is that patients arrive in her surgery asking about many types of complementary medicine. Her solution is to work with the patient to find out what is best for them. She teaches them that one of the most powerful things they can do is keep a diary of how they feel. So, for example, she introduces them to keeping a record of the pain they experience on a scale of 1 to 10. She explains that trying one type of treatment at a time is the best way to evaluate which ones work for them, rather than jumping into three or four different therapies simultaneously. In other words she shows them how to conduct trials on themselves.

In keeping with the Hippocratic Oath's central theme, a major concern for Kuhn is to minimize any potential harm to her patients. And she interprets 'harm' as both physical and economic. Complementary therapies can be expensive – especially in a country where orthodox medicine is available free at the point of use – and if patients receive no benefit from spending their money then they are suffering economic loss for no gain. Kuhn has got to know about many of the complementary practitioners in her area and while many are good she has concerns about a few. She talks with patients who are seeing those therapists and urges them to consider whether they are getting any benefit from the treatment. It is a difficult balancing act and requires thought, dedication and a great deal of commitment. It can also be time-consuming, which is a challenge when the pressure on doctors to see more patients is growing. All the same, Dr Kuhn has found a way of incorporating complementary therapies alongside her practice that works for her and her patients.

Doctors like Kuhn and Cummings make a case for adopting an open-minded approach to complementary and alternative therapies from a western medical point of view. They are persuasive because their advocacy goes against their training. And they are not alone. Legions of others hold similar views. How else to explain the 30,000 European doctors who are trained homeopaths, or the more than 20 universities in the USA offering Integrated Medicine courses, including Harvard Medical School, or the 30% of government-funded health care trusts in the UK offering some form of complementary therapy?

○

On 20 September 1997 the respected medical journal *The Lancet* published a major round up of clinical trials of homeopathy. This type of work, known as a meta-analysis, attempts to combine data from lots of different studies to see what larger conclusions can be reached. Seven researchers, based in Germany and the USA, examined 186 homeopathy experiments and decided that 89 had sufficiently good data, gathered from well-designed clinical trials, to include in their analysis. They reached a surprising conclusion:

> The results of our meta-analysis are not compatible with the hypothesis that the clinical effects of homoeopathy are completely due to placebo. However, we found insufficient evidence from these studies that homoeopathy is clearly efficacious for any single clinical condition. Further research on homoeopathy is warranted provided it is rigorous and systematic.

The language is moderate but the message is clear: the researchers reckoned that homeopathy is more effective than placebo in some instances. They could not say that it was effective for one condition as opposed to another, since the clinical trials they studied were disparate. They called for more research. Would they have made that recommendation had homeopathy been shown to be bunk?

Predictably this caused a bit of a stir. Here was one of the most respected medical journals in the world apparently giving credence to homeopathy. Debate rumbles on about how good an analysis it was, but it has never been retracted.

The body of research in support of homeopathy continues to grow. A more recent literature review, conducted by Dr Robert Mathie, Research Development Advisor of the British Homeopathic Association, appeared in the journal *Homeopathy* in 2003. Mathie found that homeopathy is effective for childhood diarrhoea, fibrositis, hay fever, influenza, pain, side effects of radio- or chemotherapy, sprains and upper respiratory tract infection. He also concluded that homeopathy is unlikely to help headache, stroke or warts.

Good quality research exists for other complementary therapies. A couple more merit a closer look. The first is a study of acupuncture published in another respected publication, the *British Medical Journal*, on 30 June 2001.

This was a single study, not a combination of many trials, of acupuncture for neck pain. The starting point for the research is that there was little evidence for the effectiveness of the conventional medical interventions, which included painkillers, massage, physiotherapy and exercise. As the introduction puts it, 'Current treatment increasingly includes complementary methods, of which acupuncture is one of the most common. There is, however, a lack of evidence to support acupuncture as an effective treatment for chronic neck pain'.

A total of 177 patients were followed over three years. They received either massage, acupuncture or a form of sham acupuncture using a laser to give an impression that something had happened. The study concluded that acupuncture was a safe and effective treatment for neck pain and that particular groups of patients seemed to benefit more than others.

Also gaining ground is the Bowen Technique, a form of contact therapy that relies on soft pressure applied to certain key points. Developed by Australian Tom Bowen in the first half of the 20th century, it is based loosely on mainstream physiology, but its philosophy is complementary. Bowen practitioners talk about 'drawing the body's attention to the problem' and 'encouraging the body to heal itself'.

The Bowen Technique is often offered for 'frozen shoulder' or adhesive capsulitis. This afflicts roughly 3% of people and usually strikes those over 40, affecting slightly more women than men. Its medical description is 'a major swelling in the layer of cartilage that lines the shoulder joint which dramatically restricts the movement of the joint'. It can be painful and makes simple everyday actions such as reaching up to a cupboard or putting on a coat very difficult. It normally clears up of its own accord in anything from 18 months to four years. Doctors prescribe painkillers, steroid injections and physiotherapy – none of which is particularly effective – and as a last resort, surgery.

An alternative is around five trips to a Bowen therapist. The therapist makes a few gentle rolling movements or soft rubs over the area with his or her fingers, or more usually thumbs. It is far gentler than most massage and totally non-invasive.

Bernadette Carter is Professor of Children's Nursing at the University of Central Lancashire. She had a persistent soreness in her ankle and was referred for Bowen therapy by her GP. She was very surprised that such a mild intervention helped and decided to investigate further. She discovered that Bowen was also popular for frozen shoulder, and decided to do some research of her own.

Professor Carter recruited 20 patients with frozen shoulder and started by measuring the mobility they had in their affected joints and asking them how much pain they experienced. Her team measured active mobility by asking patients to move their arms and looked at passive mobility by themselves gently moving patients' arms. Patients also completed the McGill Pain Questionnaire (a commonly used method of assessing pain), grading how much pain they experienced in their affected shoulder on a scale of 1 to 10, where 1 is mild and 10 is unbearable. These are well-established techniques. Each patient was then given up to five sessions of Bowen Technique frozen shoulder moves. This was not a randomized trial – it was a pilot study to see whether or not further investigation was warranted. The control for the experiment was that patients' frozen shoulders were compared with their unimpaired sides.

All showed some improvement, Carter's group concluded. Seventy per cent regained normal mobility by the end of the study – an improvement that persisted after the last Bowen treatment. Professor Carter published her pilot study in the journal *Complementary Therapies in Medicine* in December 2001. The only conclusion she offered was that Bowen Technique for frozen shoulder warranted further research.

'Warrants further research' is common to the early development of most medical treatments. If funding is forthcoming, larger, more controlled studies evaluate the treatment more carefully. The Bowen Technique has taken one of the essential first steps towards to becoming an evidence-based treatment for frozen shoulder.

○

The first time many people come across complementary therapy is in the shape of a story. 'I was desperate and didn't know what to do and then I discovered...'. The first person account is the lifeblood of women's magazines – pick any one at random from the news stand and the chances are it will have at least one tale of triumph over the odds. The column I write for *The Times* newspaper gives a scientific perspective on just this kind of story in a series entitled 'It Worked For Me'.

Testimonials to different treatments are perhaps most prevalent on the Internet. Just try typing the name of a particular therapy plus 'testimonial' into a search engine and you will be inundated with stuff like: 'I arrived at the clinic barely able to walk... I was very skeptical... as I lay on the treatment table after Mark had administered a few seemingly innocuous touches, I distinctly felt the pain start to move down the side of my body... I was able to walk to the station and take a train home'. Or 'If you haven't tried Reflexology and Holistic Medicine I'd highly recommend it. I know it's been a godsend for me'. Or 'I feel a huge difference (surprisingly to me)... it's funny... I feel more understanding or not so quick with a temper... AND I dont get tired easily at night... normally I would feel lazy and tired when it gets dark outside... but I've been going to bed late and waking up at a decent hour and I feel great...'.

Storytelling is probably as old as humanity itself: from Greek myths to tales of the court of King Arthur, from Grimm's fairytales to the legends of the Ik people of East Africa. Narrative is a form that holds our attention. And not simply because of the sudden twist or the cautionary tale, the dose of uplifting heroism or the wondrous, almost miraculous, dénouement. More than that, narrative enables us to share, to learn, to reflect and to empathize.

Small wonder then that stories can often be the starting point for discovery. It was anecdotal evidence that individuals who contracted cowpox did not catch smallpox that encouraged Edward Jenner to experiment with cowpox vaccinations. It was anecdotal evidence that led Dr John Snow to discover the cause of an outbreak of cholera in London in 1854. He followed up stories of who had contracted the disease and who had not and worked out that all those infected had drunk from the same pump. As soon as the pump was disabled, the epidemic died out. The combination of anecdotal evidence and Snow's hunches about the disease undoubtedly saved many lives.

○

Hannah Mackay is a shiatsu practitioner with a PhD in psychotherapy and has been researching shiatsu for a number of years. We talked primarily about her research, but at the end of our conversation I asked her why she practises and what she gets out of it. She told me of her success with neck and shoulder pain, something she feels skilled at and can get good results for. She told me of some of her ability to induce labour in overdue pregnant women. Then she became a little more thoughtful and moved off in a different direction. Instead of listing a number of benefits or enjoyments she began to tell me about one of her clients, 'X'.

X arrived at Mackay's practice complaining of back pain and sciatica – pains in the legs that can be extremely severe. Hannah began to treat X and almost immediately became aware that her client was 'very closed down'. Her diagnosis was that her client's body was unable to heal itself because it was 'stuck'. So she began to work on the patient's back and legs to 'open up' the body and allow it to move more freely. This did produce some reduction in the pain quite quickly, so when X returned for a second session she continued in the same vein. Once again the client reported improvement. On the third visit Hannah concentrated her efforts on manipulating meridian points 'to boost the client's energy' in the back and legs. In doing so she says she 'became aware of something frozen deep inside' and had a strong sense it was about X being bullied as a child. She had a feeling that her client didn't want to put her leg on the floor because she didn't trust it, she didn't feel safe. At the end of the session she told the client her suspicions. The client responded by telling Hannah about a trip they were due to take with their father who had abandoned them as a child. The conversation prompted further recollections, including the realization that the pains had started shortly after their father had made contact after years of absence. According to Mackay, X went away considerably better and had also gained an insight into the concerns they had about going away with their long lost father.

The point here is not the content of the story. It's the way Mackay told it. She repeated herself, hesitated from time to time and seemed to be trying to put into words something that was hard to describe. She referred to the case as a complex one and was trying to explain her experience of talking with her client about their concerns and trying to illustrate the experience of what it was that came to her as she worked on their body. More than once she described the experience as 'interesting' or 'fascinating'. Her recounting of the story reminded me of a young child relating a new discovery or an adult describing a very profound and possibly unexpected experience. She sought to convey that the engagement with her client was vividly real and was experienced on a number of levels. And she seemed to have gained a great deal from the interaction.

This depth of interaction is something that virtually all complementary and alternative therapists emphasize. Its more than just spending an hour with their clients as opposed to the 10 minutes or so available to a harried general practitioner. It is also about engaging and giving of themselves. It's evidently very important and rewarding, and is regularly described as vital to what they do. Many researchers investigating complementary medicine suspect that this deep connection with clients is part of the reason that the therapists get results.

Another noteworthy aspect of some complementary therapies is that they often require practitioners to receive regular treatment themselves, particularly while training. Mackay had to take shiatsu from another therapist in order to become registered with the Shiatsu Society. This is also the case for psychotherapists, most particularly psychoanalysts, and Alexander Technique teachers. It's an interesting requirement for a healer to have to undergo the treatment they dispense. Maybe this routine helps practitioners to maintain a strong empathy with their patients.

Mackay's testimony is also an exemplar of the way complementary therapies link physical ailments – in this case back pain and sciatica – with emotional states.

A central philosophy of Samuel Hahnemann, the inventor of modern homeopathy, was that you treat the patient, not the disease. A homeopathic consultation will include details about the patient's life, habits, likes and dislikes. Is the patient an owl or a lark? Does the patient like spicy or bland food? Do they prefer hot weather to cold, and so on. The homeopath then prescribes a remedy based on the symptoms with which the patient presents as well as on their temperament and personality. This can result in two individuals with apparently identical sets of symptoms being prescribed different remedies. Furthermore, as the condition progresses and the symptoms change, the homeopath may prescribe a new remedy. Some are pretty much universal – arnica is regularly prescribed for shock and bruising – but others are only used for particular personality types. The book *Homeopathic Medicine at Home* by Maesimund Panos and Jane Heimlich lists 19 different remedies for a cold. Aconite, for example, is described as 'better for people who are frightened or restless at night' whereas mercurius is recommended for those who have a 'coated tongue with bad odour from mouth' or are 'very thirsty, even though mouth is moist'.

This concern with treating the patient not the disease is reflected in the more mainstream medical literature as well. An editorial on integrated medicine published in the *British Medical Journal* in January 2001 was subtitled 'Imbues orthodox medicine with the values of complementary medicine'. It said 'Integrated medicine has a larger meaning and mission, its focus being on health and healing rather than disease and treatment. It views patients as whole people with minds and spirits as well as bodies and includes these dimensions into diagnosis and treatment'. It goes on to caution that orthodox medicine has promoted technological solutions to ill health over holism and simple techniques such as changing diet and relaxation exercises. The inclusion of this in such a prestigious journal is not a sign that mainstream medicine is coming around wholeheartedly to the viewpoint of complementary practitioners. Rather, it draws attention to the different approaches of the two camps. The orthodox medical view is that the body is a machine that can be cured by fixing its parts; the complementary or alternative view is that mind and body are linked and must be considered together.

A separation of mind and body in modern medicine is a source of frustration to some of the people who work within both complementary and allopathic fields. David Peters is a conventionally trained doctor, complementary practitioner and Professor of Integrated Healthcare at the School of Integrated Health at the University of Westminster in London. Paul Dieppe is also a conventionally trained doctor, a rheumatologist of international renown and Professor of Health Services Research at the University of Bristol in the West of England. Both talk passionately about the ideas of the philosopher René Descartes.

Between 1628 and 1649, Descartes wrote a series of texts laying out his thesis that the mind and the body are separate entities. They can interact and influence each other, he argued, but the conscious and the mechanical are distinct. These ideas laid the ground for the work of many later philosophers, including Spinoza and Leibnitz. Dieppe laments that the legacy of Descartes' approach is mechanistic medicine. This ignores, he says, the fact that our state of mind affects our health and vice versa. Likewise, Peters respects Descartes' radical thinking but says he is yesterday's man. Science, as he puts it, moves on. Dualism, scientific truth and objectivity have been supplanted by entanglement, uncertainty and relativity. The culture of scientific medicine needs to catch up. In short, complementary medicine majors in two things. The relationship between the patient and the practitioner and treating the entire person and not just a set of symptoms.

Clearly, the bulk of evidence in favour of complementary therapies currently takes the form of subjective, first-hand accounts – though the body of scientifically respectable evidence is growing. The practitioners continue to practice because they judge that their clients benefit. By and large therapists are looking for pragmatic, empirical evidence. The absence of 'hard' data clearly troubles some of them, but it does not drive them from their chosen path. The critics, on the other hand, criticize with the full weight of modern scientific understanding behind them. That body of knowledge is large and powerful and makes a good – but not unassailable – case for the ineffectiveness of many complementary and alternative medicines.

4

the critics

There is so much about complementary or alternative therapies that is open to criticism, from the frankly crackpot ideas about how the therapies work to the insistence of some therapists that what they do cannot be measured. Our modern understanding of science demonstrates that many complementary therapies cannot possibly work as described. This idea, know as prior implausibility, is used by some scientists to argue that there is little point in even attempting to investigate them.

Lets start with homeopathy. It is sitting there with a great big 'Kick me!' sign stuck to its back – an invitation too good to miss. Modern homeopathy was developed towards the end of the 18th century by Dr Samuel Hahnemann. He based it on the alchemists' principle that 'like cures like': a small amount of a substance that causes a condition can also treat it. For example, a tincture of dandelion is supposed to cure bed-wetting. Dandelion is a diuretic – hence its French name *pis en lit*. So any child suffering from bed-wetting will be cured, the argument goes, by taking a small amount of dandelion before they go to sleep. It's a pretty simple idea and one that has, perhaps, a smattering of real-world inspiration. Chemicals that can kill are often used to cure. Eat a handful of foxglove flowers and your heart could fail, yet taking around one four thousandth of a gramme a day of digoxin – the drug extracted from foxgloves – can help keep you alive after a heart attack. In this instance – not necessarily in general – a large amount kills and a small amount cures. Not too far from homeopathy, you might say. The main problem comes with what is considered 'a small amount'. This is a very major difficulty.

Homeopathic remedies are prepared from 'tinctures'. These tinctures are made by steeping a dried plant, sea shell, or magnesium sulphate, say, in alcohol for several weeks. Tinctures are not given to patients. Homeopaths dilute them in a large amount of water or a water–alcohol mix. They shake the mixture vigorously, then dilute that, then shake and dilute as before. The number of dilutions varies depending on the remedy. A typical preparation might be a series of six dilutions of one part in one hundred, known as '6C' and one thousand billion times more dilute than the original tincture. The resulting liquid or pills – made by spraying sugar tablets with the mixture and allowing them to dry – are the 'remedies'.

The more dilute the remedies the more powerful they are considered to be. So a 6C remedy that is a 1,000,000,000,000 (1 followed by 12 zeros) fold dilution is considered less potent than a 30C remedy which has been diluted 1,000,000,000, 000,000,000,000,000,000,000,000,000,000,000,000,000,000, 000,000,000 times (1 followed by 60 zeros). Allegedly, further dilutions increase the 'potency' even more.

This is palpable nonsense. A little maths reveals why. One drop of the original tincture contains of the order of 1000,000, 000,000,000,000,000 molecules of any active chemical. After a single 1 to 100 dilution, a 1C dilution, that will come down to around 10,000,000,000,000,000,000 molecules, which is just lopping off two zeros from the end of that very long number above. Each further dilution will remove another two zeros. So by the time the remedy has reached the 6C stage there will be 1,000,000,000 molecules of the original active ingredient left. This is a huge number, but if that active ingredient were aspirin it would represent around a thousand billionth of a gram, a speck so small you'd need an electron microscope to see it.

Remember, though, that 6C is not considered a particularly potent homeopathic remedy. To get it up to high potency, say 30C, many further dilutions occur. At 10 dilutions, 10C, there is just one molecule of the original material remaining. Do that a further 20 times to reach the 30C and there can be no traces left. Many of the highly dilute 'remedies' that homeopaths use have no active ingredients in them. They may be labelled

Pulsatilla Nigricans or *Arnica Montana*, but they are simply water. According to every scrap of scientific knowledge of pharmaceuticals, the only effect that can possibly follow from a homeopathic treatment is the placebo effect. Homeopathic remedies can no more cure kidney disease in a cat than can a glass of tap water.

That's not quite the end of the story. Homeopaths have an answer to this charge: 'water memory'. Shaking is vital, they explain, because it encourages water to 'remember' the shape of the chemicals dissolved in it. It is this memory, they posit, that has the therapeutic effect. Without the shaking, preparations are useless, they report, 'proving' that water has memory.

Water is a fluid. Fluids' atoms or molecules are randomly arranged, unlike the regular patterns in most solids. What's more, fluid molecules are in constant motion. Take two snapshots a split second apart and you get totally different but equally random pictures. Things change, though, when you dissolve something in a fluid: say some table salt, sodium chloride, in water. The water molecules arrange themselves into a sort of cage around the dissolved sodium and chloride ions.

When you remove the dissolved substance this order collapses into disorder – the neat little cage of water molecules rapidly disperses. Without the dissolved chemical to surround, the water molecules become part of the amorphous soup of the fluid again. This is driven by one of the fundamental laws of physics: the Second Law of Thermodynamics. This states that 'the entropy of the Universe increases'. Entropy is the amount of disorder in a system. A stack of bricks is low in entropy; pushing it over increases its entropy. Stacking the bricks again requires effort. Lowering the entropy of a system, in other words, takes energy.

So, to go back to the water cage surrounding the table salt. The cage is formed around the salt molecule and needs that template for it to exist, just as a sumptuous designer dress collapses into a shapeless mess as soon as the wearer takes it off. The entropy of the dress or the water cage increases when either loses its shape and it takes energy to reconstruct them. A water cage will no more retain its shape in the absence of a dissolved molecule than will a dress stand upright in the wardrobe without any support. To do so would contravene the Second Law of Thermodynamics, and no one has ever found any circumstances under which that happens. For something to do so – such as water having memory – would mean the most staggering rewrite of our understanding of, well, everything.

Advocates of homeopathy have one more card up their sleeves: the shaking. The frantic agitation at each round of dilution, they suggest, supplies the energy that keeps the cages of water together, remembering the shape of the molecules they once contained. There's a hitch, of course.

Shaking a solution does impart more energy to it. Indeed you can heat up a container of liquid by shaking it rapidly. It's a slow process: shaking a mug full of water twice a second for a minute would heat it up a mere fraction of a degree. The sides of the vessel hit the atoms or molecules in the liquid, speeding them up. Heat is just a measure of how fast molecules are moving, or more accurately it is a measure of the energy they have due to their movement – their kinetic energy. The molecules in a cup of tea move 10 times faster than those in an ice cold gin and tonic. So when the homeopath shakes a diluted solution it does gain a tiny amount of energy and heats up a fraction of a degree. However, heating a solution in this way would also increase the chance of destroying any ghostly cages of water. As the surrounding water molecules heated up and moved faster they would bash into the cages at higher speeds, knocking them apart.

In short, homeopathy is a dead duck scientifically. It uses remedies that contain no active ingredients and a laughable principle that water has a 'memory'.

Having dispatched homeopathy, let's turn to another commonly used complementary therapy: acupuncture. According to the World Health Organization, in 1990 there were 88,000 acupuncturists in Europe, of whom 62,000 were medical doctors. Over 20 million Europeans used acupuncture. In Belgium 74% of all acupuncture treatments are carried out by doctors; in the Netherlands 47% of the general physicians use acupuncture and 90% of the pain clinics in the United Kingdom and 77% in Germany use acupuncture. In 2002 the British Medical Association surveyed 365 doctors and found that almost half of them had arranged acupuncture treatments for their patients, suggesting that the trend has continued.

Acupuncture practitioners believe that there are meridians throughout the body, channels of energy that influence our organs. When these meridians become blocked, they argue, energy, or qi, cannot flow along them, so we become sick. Inserting needles (in acupuncture) or applying pressure (in shiatsu) at specific points along these meridians, the theory goes, can stimulate the flow of qi and so cure some maladies. It is a system that was developed in China thousands of years ago and has a reputation for improving a number of conditions, including pain, eczema and some gynecological complaints, such as painful periods and pelvic inflammatory disease. There's just one problem: meridians do not, as far as we know, exist.

The human body has been dissected, examined, described and documented in fantastic detail since the time of the Ancient Greeks and Egyptians. Mountains of textbooks describe the pathways of the tiny nerves and vessels that meander beneath our skin. Individual anatomists have spent their entire working lives mapping the nervous system of the head and neck or the minutiae of the reproductive system. The picture they have built up is one based on physical structures than are common to all of us.

Perhaps the ultimate expression of human anatomy is the Visible Human Project run by the National Library of Medicine in the USA. This currently contains almost 7,000 ultra-detailed images of normal male and female cadavers. The bodies have been imaged with X-rays and Magnetic Resonance Imaging (MRI) scanners; then they were sliced into very thin sections and photographed. The male cadaver has been imaged down to a resolution of one millimetre and the female cadaver down to a third of a millimetre. The resulting avalanche of data allows researchers to build up pictures of the human body from many different angles using a combination of different types of imaging. Best of all, the data is available online to all at `http://www.nlm.nih.gov/research/visible/visible_human.html`. This project allows people to go back again and again and search through the bodies to see if they missed anything last time around. You can only dissect a cadaver once, but this virtual human body can be dissected by a computer program many times.

No dissection or imaging method to date has found traces of meridians, acupuncture points or energy channels running through the body. Anatomists have looked and every time they have found nothing. Meridians, the cornerstone of acupuncture and shiatsu, are just not there.

None of which *proves* that meridians will never be found. It might be that we simply don't know what to look for – perhaps meridians are so subtle or so obvious that we are missing them. There are recent precedents for this in anatomy: for example, in 1998 urology surgeon Helen O'Connell discovered that the human clitoris is far larger than was thought. It is actually quite a substantial organ with arms that surround the opening of the vagina and an extensive structure within the pelvic region. Until that point the clitoris was generally assumed to be just a female analogue of the penis, but considerably smaller. The male-dominated medical profession was not interested in the organ and so had ignored its anatomy. Similarly, it could be that western medicine has until quite recently had little interest in meridians and so hasn't found them yet. One of the problems with this argument is that it is impossible to prove a negative. There cannot be any conclusive proof that meridians do not exist – the closest it is possible to get is repeated failures to find them. Meridian enthusiasts will always be able to say that they might exist.

But the absence of evidence for meridians is compounded by the failure to find any trace of qi energy. A modern doctor in a well-equipped hospital has a raft of techniques with which to peer inside her patients' bodies. X-rays produce detailed snapshots. MRI scanners can produce thorough images of organs as they function. There are meters for detecting nerve signals travelling to and from our brains and ways of examining the electrical signals that our muscles generate. Ultrasound measures blood flow and, of course, monitors the growth of a baby inside the womb. These techniques pick up radio waves, electric currents, sound waves and X-rays. None of them have ever shown any traces of energy travelling along paths – other than our nerves – inside our bodies.

Once again it could be that qi is invisible to all our current technology, or it could be that we are looking in the wrong place at the wrong time or in the wrong way. Once again, absence of evidence is not conclusive evidence of absence.

Take all this together, though, and the obvious conclusion is that the entire thesis of meridian-based therapies is wrong. Whatever happens in an acupuncture clinic is very unlikely to be based on energy moving along meridians within the body.

There is further damning evidence for the non-existence of qi energy and the meridians it travels along – from the practice of reflexology. This is a massage technique that purports to work on the principle that all of the organs and tissues of the body are linked to particular points on the feet. The kidneys are 'connected' to the arch of the foot and the pituitary gland to the centre of the big toe and so on. By massaging the foot practitioners claim to diagnose problems elsewhere in the body and treat them. A kidney problem might be expressed in tenderness in the arch, and careful work on that spot will help improve the condition. Once again these linkages have not been found anatomically, but it gets worse. Different schools of reflexology have different maps of the foot with different linkages. It is reasonable to counter that if the links were real then the map would be universal. All doctors worldwide agree where the organs of the body reside: someone claiming the heart is in the thigh would be treated with derision. Either reflexology is nonsense or the model of different parts of the body being connected to different parts of the foot is nonsense.

○

Osteopathy and chiropractic are two widely used manipulative therapies that are increasingly offered alongside conventional physiotherapy for, among other things, lower back pain, which orthodox medicine struggles to treat successfully. Physiotherapists are happy to accept that osteopaths and chiropractors are skilled manipulators and often work in conjunction with them. Yet once again the basic model of how these treatments work is fundamentally flawed from a scientific point of view.

The creator of osteopathy is generally accepted as being Dr Andrew Still, a surgeon in the Union Army during the American Civil War. He described his discovery as the ability to cure diseases by shaking the body or manipulating the spine. In his autobiography he wrote that he could 'shake a child and stop scarlet fever, croup, diphtheria, and cure whooping cough in three days by a wring of its neck'. The basic principle behind his ideas is that diseases are largely the result of loss of structural integrity of the body; that the bones, particularly those of the spine, are out of alignment. Osteopathy is based on the belief that manipulation of the bones, muscles and tissues can realign the body and so cure disease. Largely rejected by the orthodox doctors of the time, Still set up an osteopathic medical school in 1892.

At around the same time, another American, Daniel David Palmer, was developing another therapy based on the idea that spine misalignment was responsible for medical problems. He proposed a different model from that of osteopathy – that out-of-place vertebrae affect the nerves radiating from the spinal column and cause the illness. Chiropractors are concerned with realigning the vertebrae by manipulating the spine and prompting healing in that way.

Chiropractic and osteopathy both consider medical problems to arise from joints being out of true. Physiotherapists and orthopaedic surgeons agree that the idea that joints are misaligned in the way these therapists describe is untenable. There are just no data from X-rays or body scans that support it. The possibility remains, though, that both osteopathy and chiropractic achieve their success in exactly the same way as conventional physiotherapy, based on modern anatomy and physiology.

○

There are some forms of complementary therapy that rely on what appear to be modern pieces of technology wrapped up in pseudo-scientific claims. There is, for example, a device on sale called the Aqua Detox. This claims to remove toxins from your feet as you rest them in a bath of warm, salty water while an electric current is passed through both the bath and your feet. The proof of this is purported to be that the bath starts off nice and clean and ends up murky brown, with a scum floating on the top. Its mechanism of action prompted a lively discussion in the letters pages of the *New Scientist* magazine in July 2004.

A correspondent reported trying out the device, describing how he watched salt being added to the correct amount and the current switched on. The water duly became brown and murky, but before it became too dark he observed the discolouration coming not from his feet but from the plastic casing surrounding the electrode. Intrigued, he returned to the treatment centre the following week and this time watched the machine in action without putting his feet into the water. It duly went just as brown and murky.

A second correspondent offered a chemical explanation of what was going on. An electric current passing through a solution of salt water will produce a weak solution of sodium hydroxide, and this in turn will react chemically with the oil in the skin to produce what is described as 'a soapy gunge' – a plausible explanation for the scum on the water. The brown colour can easily be explained if the electrodes were made of iron – the brown is simply rust.

Both correspondents offer speculation, not hard evidence. However, there is a well-established principle in science known as Occam's Razor: that, of two explanations on offer, the simpler is more likely to be true. In this case the simple explanation is that a combination of rusty electrodes and oily feet produce the brown scummy water. The more complex explanation is that some as-yet undiscovered mechanism is drawing toxins through the skin of the feet resulting in brown water and a detoxified body. Occam's Razor dictates that rusty electrodes have it.

Out on another equally unscientific limb is crystal healing. This is the principle of using the 'energies' of different crystals to effect 'cures'. Different properties are assigned to different crystals and they are placed on the patient at appropriate points in order to 'energize' the body or 'release energy blockages'. There is perhaps a smattering of scientific inspiration buried within crystal healing. If quartz is squeezed it produces a tiny electrical current – the piezo-electric effect. The reverse is also true: when you pass an electric current through a crystal of quartz it changes shape ever so slightly. However, the amount of electricity produced in this way is absolutely tiny, and anyway, quartz crystals are not squeezed during crystal therapy.

Even if the piezo-electric effect were having some unspecified upshot it could not explain the claims of crystal healing, as quartz is one of the very few materials with this property. Other minerals used for crystal healing, such as jade, amber or hematite, are not piezo-electric. No wonder geologists scoff.

○

Complementary medicines are often sold on the basis that they carry fewer risks than allopathic medicine. Its certainly hard to see how taking a homeopathic remedy can do any damage if there is nothing in it. Likewise, having shiatsu or reflexology sessions is not high on the list of hazardous pastimes. Acupuncture does carry some well-known risks, most obviously infection from poorly sterilized needles. This has been largely eradicated with the adoption of single-use sterile needles similar to disposable hypodermics. Needles have broken off inside patients and, among other things, punctured the spinal cord. Dr James K. Rotchford, writing in 1999 in the journal *Medical Acupuncture*, published by the American

Academy of Medical Acupuncture, reported that five deaths had been attributed to acupuncture. Nevertheless, when properly practised acupuncture carries a relatively low risk.

Other therapies also carry risks. For example, osteopathic or chiropractic manipulation can be extremely dangerous if performed on someone with a hairline fracture of the neck. The risk is low if the treatment is carried out by a properly qualified and registered practitioner, but it would not be accurate to call the treatments entirely safe.

Herbal medications can carry quite substantial risks, particularly if combined with prescription remedies. The popular herbal anti-depressant St John's Wort should not be mixed with prescription anti-depressants, as this can cause confusion, tiredness and weakness. St John's Wort also has the effect of lowering the efficacy of the contraceptive pill and so can increase the likelihood of an unplanned pregnancy. Mixing gingko biloba, sold as a memory enhancer, with blood thinning drugs can cause hemorrhage, as gingko also acts to thin the blood. Being 'natural' does not equal being safe, and that there are genuine risks from so-called natural herbal preparations and supplements.

It is probably true to say that, with the exception of some herbal remedies, the potential dangers of these therapies is less than that of many medical interventions. No complementary therapist is planning to perform surgery or has the range of potentially lethal drugs that an orthodox doctor has at their disposal. Proper training and regulation can minimize the risk and the increasing trend towards licensing of many different complementary therapies should contribute to safety.

There is, though, a way in which a therapy that has absolutely no physiological effect whatsoever can worsen health, or even kill.

This problem arises where there is a lack of communication between orthodox and complementary therapies. Many people turn to alternatives when they find allopathic medicine wanting. This is often for chronic conditions where conventional medicine has little to offer in terms of cure. Patients with diseases such as rheumatoid arthritis are prescribed drugs that they will have to take indefinitely and that, in the main, might reduce the symptoms of the disease but not actually arrest its progress. These could include painkillers or steroids, both drugs that can have powerful side-effects. The prospect of remembering to take such drugs every day for the rest of your life is not an attractive one and it is unsurprising that patients will look for alternatives. It may be that they find some complementary therapy that helps them feel better for whatever reason and tempts them to give up taking the drugs. This can be a major mistake.

Reputable complementary or alternative therapists do work alongside their patients' doctors and are very clear about their clients continuing to take their orthodox treatment, whatever it might be. However, having two arms of treatment working alongside each other produces a climate in which people move from one to the other in a way that could do them harm.

A good example of this hit the newspapers in the UK at the end of June 2004. Prince Charles, heir to the British throne and a famous advocate of complementary therapies, has lent his support to the Gerson Therapy. This is an anti-cancer regime that involves large quantities of fresh vegetable juice, a vegetarian diet, injections of liver extract and vitamin B12, and five coffee enemas a day. Cancer experts agree that diet is important in cancer and that plenty of fresh fruit and vegetables is a good thing. However, the Gerson regime also advocates giving up anti-cancer chemotherapy. These drugs can be very powerful and the side-effects deeply unpleasant, but they are used with care and expertise by cancer specialists and their track record in treating disease is well known. The Gerson Therapy has no such medically established track record and many oncologists have warned that it might be dangerous. The American Cancer Society's web site states:

> There are a number of significant problems that may develop from the use of this therapy. Serious illness and death have occurred from some of the components of the treatment, such as the coffee enemas that remove potassium from the body leading to electrolyte imbalances.... Some metabolic diets, used in combination with enemas, cause dehydration. Serious infections from poorly administered liver extracts may result.... Relying on this treatment alone, and avoiding conventional medical care, may have serious health consequences.

○

The next thing that gets critics of complementary therapies foaming at the mouth is the evidence for its efficacy – or rather, the lack of it.

Evidence for the effectiveness of a medical treatment is gathered in clinical trials. Broadly speaking a treatment is compared with either another treatment or a dummy treatment of some sort. The quality of the data obtained depends on numerous things, including the number of patients involved in the trial, whether either the patients or the doctors taking part knew who was getting what treatment, how the patients were selected, and how they were divided between the trial group and the one receiving the established or dummy procedure. It is a complicated and difficult process, of which more later. Essentially, a well-designed trial produces reliable data.

Normally, data from lots of different trials are assessed together in what is called a systematic review – a trial of trials. Systematic reviews allocate more weight to good quality trials than bad ones.

The largest and most respected library of systematic reviews is produced by the Cochrane Collaboration, an international network of experts in the analysis of clinical trial data. These reviews, known as Cochrane Reviews, are published quarterly and are available to all to read online. They are accepted as a reliable source of data about all sorts of different treatments and make a significant contribution to which drugs and interventions are used by doctors every day.

Two other sources of information about which treatments are effective for which conditions are the Internet journal *Bandolier* and a publication from the Department Complementary Medicine at the Peninsula Medical School, University of Exeter and Plymouth on the south coast of Britain, called *The Evidence So Far*. This booklet documents research conducted by Professor Ernst and his colleagues between 1993 and 2002 into a variety of complementary therapies.

It is an interesting exercise to search for data on studies into complementary medicine in these three sources.

Let's start with acupuncture, which has perhaps the best publication track record of any complementary therapy. Out of nine Cochrane Reviews for different types of acupuncture-based treatments, three are mildly positive, drawing conclusions such as:

> Overall, the existing evidence supports the value of acupuncture for the treatment of idiopathic headaches. However, the quality and amount of evidence are not fully convincing.

or:

> This review has demonstrated needle acupuncture to be of short term benefit with respect to pain, but this finding is based on the results of 2 small trials.... No benefit lasting more than 24 hours following treatment has been demonstrated.

The other six reviews all make comments such as:

> The quality of the included trials was inadequate to allow any conclusion about the efficacy of acupuncture.

or:

> There is no clear evidence that acupuncture, acupressure, laser therapy or electrostimulation are effective for smoking cessation.

Bandolier was started as a reference source for pain treatment, and as acupuncture is often used in this way the journal has many references on it. These are peppered with qualifying phrases such as:

> Perhaps the biggest problem is that these trials, as a group, have avoided the hard question of longer-term outcomes. Even if acupuncture provides short-term relief, its place in management of back pain remains unknown.

or:

> People entering trials of smoking cessation want to stop smoking. Some of them succeed. With acupuncture, no more succeed.

What of *The Evidence So Far*? It too considers acupuncture for lower back pain and concludes 'Acupuncture was shown to be superior to various control interventions, although there is insufficient evidence to state whether it is superior to placebo'. It also has a report on acupuncture to stop smoking and once again the results make poor reading for advocates. 'The results with different acupuncture techniques do not show any one particular method to be superior to control intervention'.

A similar exercise for homeopathy, spinal manipulation (chiropractic and osteopathy), reflexology and so on yields very similar results. Data tends to be low in quality and the conclusions equivocal at best. Compare this with a Cochrane Review of the use of aspirin for acute pain:

> Aspirin is an effective analgesic for acute pain of moderate to severe intensity with a clear dose-response.

The language is careful but the message is plain: aspirin stops pain. Compare this to the hedging of the reviews of acupuncture and it becomes apparent that it really does not conclusively cut it as a therapy.

Whatever advocates of complementary therapies might argue, the current evidence for their efficacy just does not stand up well to careful scrutiny.

Nonetheless, the proper application of science is already starting to bring the critics of alternative medicine closer to its advocates and to offer the prospect of health care that has the best of both approaches. The major stumbling block at the moment is the way in which evidence for a treatment's effectiveness is gathered and used, whether it is an orthodox or complementary approach.

5

the gold standard tarnished

Two crucial questions have to be asked about any medical intervention – will it improve the patient's condition and is it dangerous? If the answer to the first is yes and the second no, then it should be applied. If the answers are the other way around then clearly the treatment should not be used. Unfortunately, life, and health care, is never that straightforward. For virtually every treatment used in medical practice today the answer to these two questions tends to be 'probably' and 'not very'. Even the most common over-the-counter drugs, such as ibuprofen, do not remove every ache and pain and carry risks. Sometimes these risks are quite severe: ibuprofen can induce an attack in one in seven asthmatics and it is not possible to tell who is going to be affected.

Any intervention carries some risk and no treatment is guaranteed to work, so what patients and doctors need is a reliable measure of how effective a drug or procedure is to set against an assessment of how likely it is to cause a problem. This risk–benefit analysis is actually independent of treatment mechanism. Determining whether something is effective is not the same as understanding *how* it produces that effect. So in theory, any form of analysis that weighs up benefit against risk should be able to test any procedure, be it chemotherapy, massage or policy.

In medicine, the efficacy and safety of a treatment are assessed by clinical trial (or more normally by a series of clinical trials), and it is the data from these that determine whether a new treatment is to be accepted and used.

Giving a drug to one sick patient tells you very little about how useful a treatment it is. If the condition was getting better anyway, you could end up ascribing the improvement to the drug. If the disease was fatal, you could wrongly conclude that the drug was deadly. If nothing happened you might decide the drug was ineffective, even though without it the patient would have worsened – but you would never know.

In short, you need to be able to make comparisons with patients who did not receive the drug. In fact, a clinical trial is a collection of sophisticated compare and contrast exercises. The design, administration and analysis of a clinical trial are complex and there are a large number of researchers whose main interest is to examine and improve the way they are conducted. A badly designed clinical trial will reveal very little of value and is a waste of time and money.

Andrew Moore is the editor of *Bandolier*, the online journal of evidence-based medicine. His role is to evaluate clinical trials and assess the conclusions drawn from them in order to provide clinicians with the best information available on how to treat patients. He uses three broad criteria for assessing clinical trials – quality, size and validity. To design and execute a trial that conforms to these criteria can be demanding and not at all straightforward.

A control is crucial. It is a fundamental principle in all experimental research without which it is impossible to draw any meaningful conclusions. A simple experiment to test the effect of water on seed germination would, for example, have two boxes of seeds, one watered and one not. The dry box would be the control. A significant result would be if the watered seeds grew and the unwatered (control) seeds did not. The control was there to ensure that the seeds wouldn't have grown anyway, and that the watering was the key. The principle is exactly the same for clinical trials.

The most obvious form of controlled clinical trial compares patients who are given an experimental drug and those who are not. Straight away difficulties appear. If you ask someone to take part in a trial and then give them nothing, they know that they are part of the control group. Consequently they may not expect to get better, which in turn can influence the way they fare. They may show no improvement precisely *because* they do not expect to improve. The obvious thing is to give some patients a placebo – a mock treatment that looks or feels like the one being tested but which has no active components. Such 'placebo-controlled' experiments are commonplace.

Perhaps the first published placebo-controlled trial appeared in 1931 in the *American Review of Tuberculosis*. The study was an investigation of the drug sanocrysin for treating tuberculosis. Patients with TB were divided at random into two groups, decided by the toss of a coin, and one group was given the drug and the other distilled water. The researchers wrote: 'The patients themselves were not aware of any distinction in the treatment administered'. Therefore, when the drug turned out to have a positive effect on the patients who had received it, the study could be cited as proof of the medication's efficacy.

Placebo-controlled trials can provide good quality data, one of Andrew Moore's evaluation criteria. There are problems with them, though, both in principle and in practice.

Ethical qualms have always dogged placebo-controlled trials. But a paper published in the *New England Journal of Medicine* in 1994 reignited the debate. The authors asserted that conducting a placebo-controlled trial when an effective treatment already exists is unethical. This, they argued, was withholding treatment from a sick patient and ran counter to doctors' avowed responsibility to treat patients to the best of their ability. To support the argument they referred to the Helsinki Declaration of the World Medical Association – the international benchmark of medical ethics. One of the results of this debate was that in 2000 the World Medical Association issued an amendment to the Declaration of Helsinki – 'the Edinburgh amendment', agreed at a meeting in the Scottish capital. This strengthened the position on the inclusion of human patients in clinical trials so that 'every patient entered into a research project should be assured of the best proven prophylactic, diagnostic and therapeutic methods identified by that study'. The debate has not stopped there. For example, a paper published in 2002 in the *American Journal of Bioethics* argued that a placebo-controlled trial is justified if patients are not exposed to any serious risk and if they volunteer.

Even if circumstances allow the use of a placebo, the practical problems of finding a suitable one remains. It is very easy to come up with a placebo for a drug. An identical looking and tasting sugar pill usually does the trick. It is far more difficult to come up with a placebo for a treatment such as surgery or psychiatric counselling.

Placebo surgery does happen. A trial conducted at the Baylor College of Medicine in the United States in 2002 compared two common operations for arthritic knees with a sham surgical procedure. The two real procedures were keyhole surgery where a narrow tube was introduced into the knee joint and the surgeon inserted the instruments down that tube. The sham was simply making small incisions on the knee to mimic the those used in the real operations. The researchers found no difference between the placebo group and the other patients.

This surprising result suggested that some people might be undergoing unnecessary operations on their arthritic knees. 'We have shown that the entire driving force behind this billion dollar industry is the placebo effect' said study leader Dr Nelda Wray. 'The health care industry should rethink how to test whether surgical procedures, done purely for the relief of subjective symptoms, are more efficacious than a placebo'.

The knee operation was a relatively minor, minimally invasive procedure and the condition it was treating was not life-threatening. The patients recruited for the study were told that they might receive a sham operation and had to write on their medical charts that they understood this. This put off a significant number of potential patients: 44% of those originally identified for the trial refused to take part. This is a potential problem in any medical research, but 44% drop-out is very high.

Clearly the designers of this trial were confident that a placebo operation was ethical, as was the ethics committee that would have approved it. Every hospital or clinic where trials take place has an ethics committee that decides which studies may go ahead. It might have been a different story with something like heart bypass surgery.

Heart bypass surgery – replacing clogged arteries in a patient's chest with healthy blood vessels from their legs – can help prevent heart attacks. Hypothetically speaking, if there were an alternative form of surgery then one way to test its efficacy would be a placebo-controlled trial whereby some patients underwent the new procedure and some were cut open but then sewn up again straight away. Sham surgery would most likely constitute withholding effective treatment, and is thus extremely unlikely to be approved. A viable placebo, while being possible here, would not be ethical.

Taking this one step further, there are some surgical procedures for which it is impossible to imagine a placebo – from something as minor as remedial work for an ingrowing toenail to something life-saving, such as removing a malignant tumour. These important everyday medical procedures are simply not amenable to testing by placebo-controlled trial.

Clearly another form of control is sometimes required. The most common approach is to compare a new therapy with an established one. This sort of trial has two groups of patients with the same or similar condition; one group is treated with the established therapy and the other with the experimental one. Ethical concerns are dealt with in that every patient recruited onto the trial is given a treatment that is either known to work, or is expected to. And the problem of devising a suitable placebo falls away.

○

Think back for a moment to the 1931 TB drug trial. There was another crucial element to it: patients were divided at random into two groups, one of which received the drug and the other the placebo. This element of randomization is probably the most important feature of a clinical trial.

Randomization removes the vagaries of human individuality and allows general conclusions to be drawn. Randomization avoids the possibility that all the sickest patients end up in the control group and the less sick in the test group, an arrangement that would skew the results one way or another. It minimizes bias due to individual variation in responses to the drug. It smooths out differences in temperament, diet, lifestyle or the way in which the patients take the drugs. Professor Sir Iain Chalmers, former head of the UK Cochrane Centre, says that randomization is *the* crucial element. 'Randomization is the only thing special about the trials. When you are comparing like with like you have to randomize.... If you want to skip that you have to explain how you have ensured comparing like with like and this is the Achilles Heel of trials that do not randomize'.

Combining the two elements – randomization and a suitable control – results in the randomized controlled trial. This is the internationally acknowledged best means of testing a medical intervention of whatever sort, be it drug, surgery or physical therapy.

This is not the end of the story. A randomized controlled trial is still open to another form of bias that comes from the doctors and patients taking part in the study.

In 1784 King Louis XVI of France ordered an inquiry into the phenomenon of 'mesmerism' or 'animal magnetism'. This is the supposed ability to affect a 'universal magnetic fluid' that runs through the body and produce trances and healing as a result. The inquiry was headed by American Founding Father Benjamin Franklin, and included the noted chemist Antoine Lavoisier – famous for discovering oxygen. One of the tests was on a group of blindfolded subjects. They were either given the 'magnetism' or not and simultaneously told the truth about what was going on or lied to. The inquiry found that people only 'felt the effects of mesmerism' when they were told it was happening, regardless of what they really got. This was one of the first recorded blinded experiments – literally in this case – where the subjects did not know whether or not they were receiving treatment.

Blinding is an important part of a modern randomized clinical trial. A patient is told at the outset that they are going to receive an experimental treatment or a control treatment as ethics demands. But to ensure that their expectations do not influence how they respond, patients are not told which they are getting. This is called a single blind trial.

A second source of bias derives from the doctor administering the treatment. A doctor might react differently if he or she knew that the patient in front of them belonged to the control group, say. Therefore doctors are also often kept in the dark about which treatment they are administering. This is a double blind trial.

○

The data that emerge from a properly randomized, properly controlled, double blind clinical trial are likely to meet the first of Andrew Moore's criteria for good research. They are more likely to be good quality data.

The next criterion is quantity. A randomized controlled trial produces an average of patients' responses to the treatment. The more patients taking part in a trial, the more likely it is that the results will produce a meaningful average. It is not always possible to get a large number of suitable patients for a study –

the condition might be rare, or the treatment might only be designed for a small subset of people. Studies of physical interventions tend to be much smaller than those for pharmaceuticals. It is quite possible to recruit fifty or sixty thousand people for a drug trial, such as those involving the cholesterol-lowering statins. Performing a similar number of operations to assess different types of heart bypass surgery is, logistically, very unlikely.

The more patients in a trial the lower the potential for errors and the greater the likelihood of obtaining meaningful results, that is results with a high statistical significance. To work out the minimum number of patients needed for a meaningful trial can be fiendishly complex. Considerations include the rarity of the condition under test, how many patients may drop out or not take the drug, and how many false negative results the researchers are prepared to tolerate. Ethics comes into play too, as if the sample size is too small to produce meaningful results then patients could be put through a useless exercise.

The last of Andrew Moore's criteria is validity. This refers to the way in which the data have been analysed and the conclusions drawn. If, for example, a study shows that a new painkiller produces a 4% improvement in 3% of patients it is clearly far less efficacious than one producing a 50% improvement in 70% of patients. Even if the trial is immaculately designed, double blinded and including thousands of patients, it could still fail to be valid. There are different ways of analysing data, all open to different interpretations. And data analysis, like all the elements of a randomized controlled trial, is still an evolving discipline, and there are no hard and fast rules for doing it well.

The randomized double blind controlled trial has become the gold standard by which medical interventions are assessed, the most convincing analytical tool in clinical research. It is not, however, all powerful. There are limits to what such a trial can discover and what it can be used to investigate.

○

The first constraint is time. A typical trial might run over two or three years, within which time each patient might actually receive the treatment for between a week and six months. This may be more than enough to investigate the effects of a new drug for pneumonia or a new type of appendicitis surgery, but is too short a time to study an intervention for a chronic condition. Take, for example, smoking.

It is as good as proved that smoking causes all sorts of very serious ailments including heart disease and lung cancer. The most powerful solution is to give up, yet this is extremely difficult to test in a randomized controlled trial. The effects of smoking can take years to appear and it is totally unfeasible to run a trial that would last long enough to see whether giving up smoking really does improve health. Patients would have to either smoke or not smoke for half of their lives, and the lives of the researchers, before sufficient meaningful data could be gathered. The same problem affects drugs such as beta-blockers that people often take for the rest of their lives.

There are good ways of gathering long-term data. Sir Richard Doll famously linked smoking and lung cancer by looking back at smokers' medical records. This retrospective study examined the effects of smoking on health *after* they had appeared. A clinical trial, by and large, looks for effects *as* they appear.

Another limitation of the randomized controlled trial is that it cannot pick up rare events easily. This is simply a result of the numbers of people who need to be involved for something unusual to show up. If, for example, a drug causes a side-effect in 10% of the population, that side-effect will, in theory, be likely to turn up in a trial of just 10 patients. If the effect appears in just 1% of patients then the trial will need to include at least

100 people. When you get down to side-effects that occur in very small percentages of the population, only 1 in 50,000 say, then only a very large trial would pick this up. Such rare events are more often spotted in retrospective studies than in randomized controlled trials.

The next limitation is the fact that one randomized controlled trial is not enough. Each only asks a very specific question and that might not be the only question hanging over the condition or intervention under investigation. Take, for example, antibiotic treatment for urinary tract infections, also known as cystitis.

Cystitis can be very painful. It is found in people of all ages, particularly young children. By the age of five 1.7% of boys and 8.3% of girls have had a urinary tract infection. Antibiotics are the preferred treatment and are relatively straightforward to trial. Yet to build up a good picture of the clinical use of antibiotics in cystitis several different trials have been required. The Cochrane Collaboration currently reports that 13 valid clinical trials have looked at the optimal duration of treatment for elderly women, ranging from a single dose to two-week courses. Eleven have examined the use of the antibiotic methenamine hippurate as a way of preventing urinary tract infections. Three trials have studied long-term courses of two different antibiotics in children. Ten have examined whether short or long courses of treatment were most effective at clearing up the infection in children. In all, there have been at least 37 different trials just to explore the use of antibiotics in treating one type of infection.

Each clinical trial can be seen as one point in a join-the-dots picture. Not until all the dots have been joined up does the full picture become visible. It is not necessary to join every single one to get an idea of what the picture might be, but the more you do the greater the definition. A few good clinical trials will provide a useful outline of how a treatment works for a particular condition, but to get a detailed understanding many trials are needed.

There is an important consequence of this. Clinical trials examine one element of the treatment at a time; therefore they are very poor at looking at multi-factor interventions – more common in medicine than it might first appear. People with high cholesterol levels might be put on statins for life, but they will also be advised about diet and exercise at the same time. All are known to influence heart disease, but a clinical trial could only look at these elements in isolation. In this case it is relatively straightforward to design a trial that looks at, say, changes in diet while holding drug levels constant. But it can be problematic if a treatment is made up of a many different factors that cannot be separated.

The UK government funding body, the Medical Research Council, acknowledged this problem in a discussion document published in April 2000. Called 'A Framework for Development and Evaluation of RCTs for Complex Interventions to Improve Health', it contained a hypothetical scenario to illustrate the difficulties that 'bedevil well designed research':

> ... what is a physiotherapist's contribution to management of knee injury? The package of care to treat a knee injury may be quite straightforward and easily definable – and therefore reproducible: 'This series of exercise in this order with this frequency for this long, with the following changes at the following stages'. However, the physiotherapist may have, in addition to the exercises, a psychotherapy role in rebuilding the patient's confidence, a training role teaching their spouse how to help with care or rehabilitation, and potentially significant influence via advice on the future health behaviour of the patient. Each of these elements may be an important contribution to the effectiveness of a physiotherapy intervention. If we now

> hypothetically consider evaluating a specialist stroke unit, the physiotherapist is one potentially complex contribution in a larger and more complex combination of diverse health professionals' expertise, medications, organizational arrangements and treatment protocols that constitute the intervention of that unit.

The document suggests that observational studies could be introduced as part of the way of measuring the trial outcomes. That is, rather than just measuring changes in clinical symptoms, such as the reduction in swelling of an arthritic joint, researchers should attempt to observe how a patient behaves as a result of the study. These ideas of broadening the purview of clinical trials are central to exploring complementary medicine.

The randomized controlled trial really struggles to cope with any form of psychotherapy. Whether it is given by a psychiatrist, a clinical psychologist or someone trained specifically as a psychotherapist is immaterial. The problem is the nature of the relationship between the patient and the therapist.

Firstly, it is impossible to conduct a double blind trial with any form of therapy that involves talking. One might construct a session wherein the patient is unaware of whether they are getting psychotherapy or just conversation, but it is impossible for the therapist not to know. At best any such treatment can only be single blinded.

Secondly, psychotherapeutic treatments tend to take a long time. Clients are typically in therapy for months or years, making them difficult and expensive to track.

Thirdly, the quality of the treatment is based on many things: the experience of the therapist, the relationship between the therapist and the patient, and the preparedness of the patient to cooperate. In other words there is no standard psychotherapeutic intervention in the way that there is a standard dose of a drug. Recruiting sufficient patients gets round some of these issues, just as recruiting thousands of people onto a pharmaceutical trial will average out individual responses. But what remains is that each psychotherapeutic intervention is specific to the therapist and patient. It is a tailored treatment. What works for one may have little effect on another.

As I write, 5,685 clinical trials are under way around the world according to the *meta*Register of Controlled Trials, an international database developed and maintained by the publishing consortium Current Science Group. These trials are good at determining the efficacy of a single intervention for a single condition. The data they produce are the average responses of a group of patients. A single trial answers a single question and many are needed to build up a full picture of how effective a particular treatment is. To get reliable data large numbers of people need to be studied, the more the better. The trial has to be properly controlled, whether by comparing the intervention with a suitable placebo or with an existing therapy. Trials have to be conducted ethically and no patients must be exposed to unnecessary risk.

Double blind randomized controlled trials are poor at assessing long-term interventions. They are poor at evaluating treatments that rely on the therapist–patient relationship. They are unable to pick up rare events or easily evaluate treatments for rare conditions. It is very difficult to design them for complex interventions and for individualized treatments. These limitations are known and accepted by the clinical researchers who use them and the data that come from them.

○

So far this discussion of randomized controlled trials (RCTs) has stayed within the realm of orthodox medicine. As the most powerful and most persuasive tool in the medical researcher's armoury the RCT is also the one most commonly applied to complementary therapies. This is all well and good for some treatments: herbal remedies, for example, can be trialed like conventional pharmaceuticals.

But the majority of complementary therapies fall into the category of treatments that randomized controlled trials are poor at assessing.

Homeopathic remedies are pills, tinctures or creams and so on, and could, one might imagine, be tested in the same way as pharmaceuticals. But that is not the only aspect of the therapy. Hahnemann, the inventor of modern homeopathy, insisted that the practitioner treat the patient, not the condition. As a result, two patients visiting a homeopath with similar problems may emerge with very different remedies. It is possible to design a trial to investigate a single remedy for a single condition, but homeopathy's use of different remedies for the same condition makes it a complex intervention, and therefore difficult to measure with a randomized controlled trial.

Similarly, acupuncturists will use different acupuncture points on different occasions while treating the same patients for the same condition. Shiatsu is also a complex intervention. The points that a practitioner stimulates vary depending on the state of the patient at the time of the consultation. In the same way, chiropractic, osteopathy, the Bowen Technique and reflexology all have elements that push them into the realm of complex interventions.

Intrinsic to many complementary therapies is advice on diet and lifestyle. Traditional Chinese Medicine, on which acupuncture, reflexology and shiatsu are based, is explicit in talking about how nutrition can affect health. Foods are divided into 'hot' and 'cold' – nothing to do with their temperature. 'Cold' foods include watermelons, pears and spinach and are prescribed to 'de-toxify' the body; 'warm' foods, among them ginger and garlic, are used to 'boost energy' and 'assist the body to heal itself'. Homeopaths will advise on diet as well, but almost all insist on abstinence from coffee and peppermint, as these can interfere with the remedies. It is possible to design trials that account for dietary factors as well but it complicates matters.

But where RCTs really struggle when it comes to complementary medicine is the practitioner–patient relationship. Alternative experts spend far longer on consultations than orthodox doctors. An initial session can easily be one and a half hours long, and follow-ups normally last between half an hour and an hour. In that time the practitioner will quiz patients about their lifestyles and their psychological state as well as their health problems.

The practitioner will then tailor what they are doing depending on the response of the patient. This could be a simple 'ouch' as a masseur presses a tender spot, but often or not the feedback is more subtle. Acupuncturists are trained to be sensitive to a large number of different 'pulses' throughout the body. Shiatsu practitioners report being aware of 'energy' moving within a client's body. Whatever the debates about the origins or even the existence of these phenomena, such therapists maintain that they can only detect and work with them if they pay close attention.

The consequence of this direct interaction is that the process of healing, according to many complementary practitioners, is one of becoming aware of a patient's state and responding to it. The therapeutic relationship, therefore, is crucial to the effectiveness of most complementary therapies. And randomized controlled trials are poor at assessing complementary treatments that involve such a relationship, just as they are poor at assessing orthodox psychotherapeutic ones.

These problems are not exclusive to complementary therapies. Doctors argue for the centrality of the doctor–patient relationship in good medical practice and are sensitive to changes that threaten it. A standard consultation by a family/ general practitioner has many of the aspects that make complementary therapy sessions hard to scrutinize with randomized controlled trials. GPs vary the treatment offered on the basis of how patients responded to the last option in much the same way that homeopaths or acupuncturists do. Doctors offer advice on eating and exercise.

The fundamental difference is that, by and large, an orthodox doctor is offering a series of treatment options that are considered as individual entities and assessed as such. Complementary therapy considers all of these elements together. Splitting them apart destroys the treatment.

○

The difficulties of designing randomized controlled trials to assess complex interventions and the therapeutic relationship are not the only ones that bedevil research into complementary therapies. There is the associated and knotty problem of 'outcomes'. That is, by what result, topic, event – outcome – is the impact of the intervention under investigation measured?

The selection of outcome measure is a key part of the criticism that advocates of complementary therapies level at randomized controlled trials. 'Well, no wonder you think this therapy doesn't work', they say, 'because you are deciding its success in terms relevant only to allopathic medicine'. Yet these cavils too go beyond complementary medicine.

In the wake of fierce debates through the 1970s about the medical control of childbirth, sociologists Ann Oakley and Hilary Graham pooled the results of their separate research projects on obstetricians' and mothers' different frames of reference of the experience. They found that obstetricians see reproduction as a medical topic, with pregnancy and birth as physiological processes, and the pregnant or in-labour woman as a patient, just one of a number on their case load. Women, however, see bearing a child as a natural process integrated into other aspects of life. This was brought into stark contrast when each group identified what a successful birth meant to them. For obstetricians, a healthy baby and well mother defined a successful pregnancy outcome. For a mother the definition was more complicated. Certainly a healthy baby was of prime importance, but a satisfactory experience of the birth itself and events thereafter were often included in her definition of success. An obstetrician might consider a birth successful that a mother deemed deeply dissatisfying if for instance if she felt uncared for and just another patient on the maternity production line. Conversely, an obstetrician might help deliver a sick baby and so record it as a failure, whereas the mother might have felt cared for and supported throughout the experience and see the birth as a success.

Clinical trials are designed to look for improvements in clinical symptoms. So a trial investigating the ability of turmeric to speed wound healing would measure how fast lesions closed up. The researchers would agree a definition of what 'healed' meant, how long you would normally expect a wound to heal and what was a suitable wound to be included in the trial. If the wounds treated with turmeric healed faster than the non-dosed wounds then it would be likely that something in the spice did indeed speed healing. More accurately, the trial would have shown a correlation between applying turmeric and faster wound healing according to the criteria used in that trial. That may sound pedantic, but it is crucial. A clinical trial tests only the criteria it sets out to look for: it is not a general exploration of what happens. Long before any patients are recruited, every researcher involved has to agree what they will look for and what they will consider a success.

When alternative medicines are put under the RCT spotlight the criteria for success have usually been traditional ones, such as reduction in inflammation in joints for osteoarthritis. Many complementary therapists argue that this ignores other benefits of their medicine.

Up to half the visits to acupuncturists in Britain are for the relief of osteoarthritis. Yet a systematic review of 13 trials of acupuncture in osteoarthritis, written by Professor Edzard Ernst and published in *Bandolier*, concluded that there was no evidence that it was more effective than either sham acupuncture, which pretends to pierce the skin, or placebo acupuncture, which does so at the wrong spot.

Several trials tracked the degree of pain that patients experienced. As pain is one of the main things that osteoarthritis patients complain about, it is an obvious measure of a treatment's success. Some researchers argue that it is not that simple.

○

John Hughes is a diffident, slightly earnest PhD student. We met at a conference on Developing Research Strategies for Complementary Medicine and got chatting over lunch as he waited nervously to deliver his short talk. He is doing qualitative studies of rheumatoid arthritis patients' perceptions of acupuncture. Rheumatoid and osteoarthritis have very different causes, but both give suffers life-long joint pain and both are difficult to treat with conventional medicine. Acupuncture is often given for both conditions, despite clinical trials finding 'no evidence' that is effective in either. Hughes believes that researchers might not be asking the right questions.

Qualitative research comes from social rather than natural science and has not always been as readily accepted in medicine as quantitative research. In a nutshell, qualitative research investigates what is happening, whereas quantitative research probes how much.

Hughes is doing a qualitative study of 22 patients with rheumatoid arthritis receiving acupuncture. He is conducting a series of open-ended interviews to find out what benefits patients *feel* they get from it. The people Hughes is questioning report an improvement in depression and/or mobility. Some admit they experience little or no pain relief, but say they feel more able to go about their daily lives. This hints that were clinical trials able to measure quality of life they might show acupuncture to be as useful as anecdotal evidence suggests.

The problem of outcomes in rheumatoid arthritis trials has been officially recognized by the OMERACT initiative. This is an international informal network of individuals, working groups and gatherings interested in rheumatology outcome measurement. The acronym originally stood for Outcome Measures in Rheumatoid Arthritis Clinical Trials, but they have subsequently dropped the arthritis clinical trials and changed rheumatoid to rheumatology. The outcome measures that the group issues are the *de facto* global standard for assessing clinical trials. The sixth OMERACT meeting in 2002 featured a major session on patient perspectives and recognized that these subjective experiences are not properly considered in rheumatoid arthritis trials. As a result, OMERACT is actively researching ways in which the outcome measurements themselves may be looked at in clinical trials of treatments for rheumatoid arthritis.

This type of research is relatively new in complementary therapy, and to date there are few completed studies. One such, published in January 2003 was commissioned by the European Shiatsu Foundation and entitled 'The Experience and Effects of Shiatsu: Findings from a Two Country Exploratory Study'. This was conducted by Andrew Long and Hannah Mackay of the University of Salford. It only looked at a small number of people: 15 patients and 16 practitioners in Germany and the UK.

The study had two aims: to 'uncover client and practitioner perceptions of the experience and effects of shiatsu' and to 'develop a protocol for the undertaking of cross-European cohort study of shiatsu clients'. Like Hughes' work on acupuncture it was attempting to lay the ground for a larger study by establishing how to measure the effectiveness of the therapy.

The researchers concluded that clients experienced a number of short-, medium- and long-term effects that they (the clients) felt to be beneficial. Long and Mackay reported the importance of clients having confidence in the practitioner and that the effects of shiatsu changed over time. They state that the study achieved its aims and, crucially, that it provides a 'base on which to design appropriate measuring tools for a wider study with larger numbers'.

Both of these studies are attempting to ask questions about what types of outcomes can be expected from two different types of complementary therapy. Neither is trying to find out whether the treatments are effective, just how better to define what they do. This is analogous to obstetricians asking pregnant mothers what they want from the birth of their child. The benefits of these types of study will not be limited to complementary medicine, but might also shed light on the experience of being treated by an orthodox doctor.

○

There is another, if not widely respected, form of clinical trial with which complementary medicine researchers are attempting to study individualized treatments and reactions: the *n* of 1 trial.

The symbol *n* in a clinical trial represents the number of patients involved; conventionally, the larger the number, the more reliable the results. One of the commonest criticisms of any trial is that the sample size was small – '*n* was low' – and so few reliable conclusions can be drawn. Flying in the face of this logic, *n* of 1 trials are randomized clinical trials with just one patient.

A single patient with a condition is given a number, usually two, of different treatments or one treatment and a placebo over a period of time. Typically there are three pairs of treatment periods. The point of the trial is to determine which of the two treatments is better suited to that particular patient. It is not so different from a doctor trying out different treatments for a patient, just more formalized.

In a guest editorial in the December 2003 edition of the journal *Complementary Therapies in Medicine*, chartered statisticians Anna Hart and Christopher Sutton discuss *n* of 1 trials. For an *n* of 1 trial to work, they say, the treatment under investigation has to be fast to act and fast to stop working when ceased. If it is slow to act and slow to disappear its effects might spill into the trial period of the other treatment. Unfortunately, complementary and alternative medicines are often slow acting. This constraint can also rule out treatments that include lifestyle changes, such as major alterations in diet or exercise habits.

In the same edition of the journal a research paper combines the results of 24 *n* of 1 trials – on 24 different patients – studying valerian root and chronic insomnia. The researchers found no evidence that valerian does help insomniacs sleep, within the constraints of the trial. The important point here is not the result of the trial but the way it was conducted. As the editorial points out, combining 24 *n* of 1 trials is unusual and raises many methodological questions. However, in the way of science, the paper was published so that others could examine and criticize its methods and conclusions. It may be that *n* of 1 trials do not offer a solution to some or any of the problems of assessing complementary therapies, but until they are tried and scrutinized it is impossible to say.

○

There is little doubt that the effects of complementary and alternative medicines are often subtle – if not there would be little debate. As a consequence it is proving difficult for researchers to develop ways of testing them. I have outlined the intrinsic flaws of the randomized controlled trial – medical research's gold standard – and some of the ways in which researchers are trying to adapt it to the study of unorthodox medicines. This is difficult and is largely at the experimental stage. Debates still rage over the right questions to ask, the right statistics to use, the right trial designs and the correct outcomes. But that is the nature of research, if we had the answers there would be no need to investigate.

In November 2000 the UK Government House of Lords Select Committee published a wide-ranging report on complementary and alternative medicine. One of the appendices noted: 'Concerns over RCTs distorting a therapy or disguising its efficacy are not the unique concerns of CAM practitioners. Vincent & Furnham suggest that as attempts to apply the RCT to a wider and wider range of treatments have occurred, more and more problems have been uncovered. They list 10 such problems.... All these methodological issues apply to both conventional and CAM treatment trials. Therefore CAM is not necessarily a special case requiring radically new methodologies'.

There is an implicit coda to this statement. Research methods that provide answers for complementary therapies will have wider applications in medical research.

Much of the research into complementary and alternative medicines is actually research into how to do research. It is about developing the correct tools by adapting the ones that are available and inventing new ones. Another way of looking at it is that as medical research is testing alternative therapies, alternative therapies are testing medical research.

6

measuring the unmeasurable

Let's go back to John Hughes and his research on patients with rheumatoid arthritis. Hughes had noticed a discrepancy between clinical trials suggesting that acupuncture is ineffective and social survey data reporting that patients were happy with it. He began by interviewing acupuncturists to get an idea of what results they expected from treatment. Although he discovered a difference between traditional and Western acupuncturists, both were concerned to alleviate symptoms and to help patients to live with the disease. Hughes asked patients what they felt acupuncture had done for them. These two sets of interviews suggested to Hughes that the outcome measures typically used in clinical trials of acupuncture for rheumatoid arthritis are missing something.

Hughes describes the interviewing he uses as in-depth and semi-structured. He audio-tape records his interviews, transcribes them and analyzes them using grounded theory, a research approach introduced in 1967 by American sociologists Barney Glaser and Anselm Strauss. This is just one of the types of qualitative research beginning to be used to study complementary therapies.

Andrew Vickers is Assistant Attending Research Methodologist at the Memorial Sloan-Kettering Cancer Centre in New York. He has written extensively on complementary medicine research. In 1996 he published a commentary in *Complementary Therapies in Medicine* on a report which appeared the previous year from the UK's Nuffield Institute of Health, entitled 'Researching and evaluating complementary therapies: the state of the debate'. Vickers identified several 'myths' that the report had revived:

> It is claimed that 'more quantitative methods' are 'preferred by orthodox medicine' and 'more qualitative methods' are 'associated with complementary therapies'. No evidence is presented to support this claim. Not one

> example of qualitative research in complementary medicine is quoted in the report. Though the case studies which are found in many complementary journals might be considered qualitative in nature, it is doubtful that any meet even a small number of the methodological criteria used to assess the rigour of such research.
>
> The authors also complain about the low status of qualitative research in the hierarchy of evidence and recommend that 'research ... give greater credence to the use of qualitative methods.' It is not generally thought that qualitative research can, by itself, assess questions of effectiveness. Among other roles, qualitative methods can play an important part in determining the questions asked in quantitative trials and in helping to implement their results. However, they are not usually thought to produce direct evidence of effectiveness. The authors of the Nuffield report do not explain how qualitative methods could play such a role in complementary medicine.

In other words, qualitative research is not a cure-all.

All the same, there are calls from various quarters for qualitative work to take some sort of place in medical research, even if it cannot directly measure how effective a treatment is. The House of Lords' Science and Technology Select Committee on Complementary and Alternative Medicine is cautiously keen, although the report's discussion of the matter suggests that their Lordships, like many others, still have an uncertain grasp of quite what qualitative research actually *is*.

The discussion document from the Medical Research Council suggests taking a step-wise approach. They outline five phases of what they call a 'continuum of increasing evidence'. The main trial only happens at phase 4 and is preceded by work which includes teasing out elements of the intervention. Among the techniques available for this groundwork, the authors (a group of health service researchers and sociologists) list, a little inconsistently, various methods of qualitative data collection, including group interviews (focus groups), individual in-depth interviews, observational (ethnographic) research, preliminary surveys and organizational case studies.

'Qualitative research' the authors explain 'may save the researchers from making inappropriate assumptions as they proceed to design the next stage of the work.'

So even when tackling randomized controlled trials to evaluate complex interventions, qualitative research is recommended as a useful preparation.

A slightly different angle appears in a paper published in the *British Medical Journal* – 'Discrepancy between patients' assessment of outcome: qualitative study nested within a randomized controlled trial' – in 2003. This features an RCT of a suite of physiotherapy treatments and the provision of advice for patients suffering pain in the knees because of osteoarthritis.

The authors carried out in-depth interviews after treatment but before the main follow-up stage. They employed an experienced interviewer who used a checklist of topics to ensure that they covered the same ground with all 20 patients. Each interview was audio-tape recorded, transcribed in full and then analyzed independently by two researchers. To make sure these analyses were fair the researchers were blind to the answers that patients had provided to the quantitative questionnaires. Both quantitative and qualitative data included patients' assessment of their condition in terms of (1) improvement, (2) worsening or (3) no change in the pain that they felt or the extent to which their activities were restricted.

The authors describe their results as 'disquieting'. They found less than 50% agreement between the quantitative questionnaires and the qualitative interviews. They attribute this discrepancy to the circumstances in which the data were gathered. The quantitative data were collected in the presence of a doctor in the clinic where the trial was based. The qualitative data, by contrast, were collected by an interviewer who is not a health professional and in the patients' own homes – i.e. *their* 'territory'. This is a reminder that the context in which people say things affects what they say. The point, stress the authors, is that it is essential to get a good understanding of the variety of what the patient feels and thinks about a treatment when trying to find out whether an intervention is effective. And to do that the patients' viewpoints have to be studied in a way that is considerably more subtle than a quick questionnaire will allow.

Here qualitative research is *part of* a study. It is 'nested' into the trial rather than used as a stage in developing its design and is being put to work as a sort of a double check on the usefulness of quantitative data.

Calls for the inclusion of qualitative work are also, of course, echoed in alternative medicine research. In 2002 spiritual healer Su Mason and her colleagues Philip Tovey and Andrew Long, social researchers in health, published an article in the *British Medical Journal*'s 'Education and Debate' section. They point to some of the limitations of randomized controlled trials for complementary therapies and note that evaluations of complementary therapies need to include adequate attention to 'holism'; 'the intent to heal, non-judgemental listening... the healing environment, [and] users' expectations or attitudes' can and should be investigated by qualitative studies, they urge. They end their piece with a significant remark, almost a throw-away, but one that echoes a refrain that complementary therapists and their patients return to again and again: 'from the user's perspective, it is the beneficial effect itself that matters not how it was brought about'.

Complementary practitioners are not the only ones calling for qualitative research into complementary medicine. Qualitative studies, wrote Professor Bernadette Carter in the journal

Complementary Therapies in Nursing and Midwifery, can 'help researchers understand the meaning, beliefs and expectation that patients ascribe to CAM interventions'.

Randomized controlled trials, Carter notes involve a 'scientific, objective and detached' attitude to a study's outcomes. For complementary practitioners such an attitude is the opposite of their normal approach, which is to be 'subjective, engaged, helping, holistic'. On these grounds, she argues, it is inappropriate to introduce a study design which means that a practitioner has to go against their philosophy. Underlining her point, she also notes that patients have high expectations of complementary treatments. Her solution is to ensure that the research design reflects these features through qualitative research that explores the relationship between practitioner and patient. In her study of Bowen Therapy for frozen shoulder, she accommodated the therapists' desire for dependable evidence that would both be understood by allopathic practitioners and 'capture the "essence" of Bowen Therapy' for this group of clients. Complementary medicine needs, she says, to 'compete in the tough world that is allopathic medicine, they still have to be able to play the same games, only play them better'. It may be a challenge, but the line to pursue, she believes, is 'to fuse both qualitative and quantitative approaches'.

The reason that people have been calling for qualitative research is that it is relatively new as applied to medicine. However, a detailed 1998 review of the literature on 'Qualitative research methods in health technology assessment' by health care sociologist Elizabeth Murphy and her colleagues revealed that they are not a recent invention, dating back some 2,000 years. Hardly a new kid on the block!

Of the several types of qualitative research two in particular have been applied to studying complementary medicine. These stand in sharp contrast to one another. The first frequently uses in-depth, semi-structured interviewing along the lines of Hughes' study. Such interviewing is also sometimes called 'open-ended' and occasionally 'unstructured'. At times this approach may also involve observation. This divides into participant-observation and non-participant. In the former the researcher takes on a role relevant to the setting to be investigated, for example that of a clerical assistant or a hospital porter. Or it could mean just moving to live alongside the community under study. In non-participant observation the researcher avoids active engagement in any capacity other than observer. The second main type of qualitative data collection is conversation analysis, of which more shortly.

Qualitative interviewing is more like a conversation than a formal question and answer session with the interviewees encouraged to take the lead and interpret the topics as *they* understand them. Good qualitative interviewers do all they can to avoid imposing their own assumptions, offering instead maximum 'space' to the interviewees to express theirs. In particular, qualitative interviewing offers the interviewees the chance to raise issues that may not have occurred to the researchers. These discussions are usually audio-tape recorded (occasionally video-taped) and then transcribed for analysis.

These exchanges can be one on one. There can be two interviewers to one interviewee, with possibly a silent note-taking observer. Or they can be conducted in a focus group, a technique most often thought of as a market research tool but which is becoming more common in academic research.

Ursula Sharma, Director of the University of Derby's Centre for Social Research, wrote in her book *Complementary Medicine Today: Practitioner and Patients* about an unusual study she did of the use of complementary medicine in and around Stoke-on-Trent, an English town in the Midlands, away from the usual concentration on major metropolitan areas. She carried out in-depth interviews with people who used complementary medicine. Qualitative data, Sharma says, are 'probably more useful than large-scale survey data for this purpose, since they are better adapted to the study of process'. She acknowledges that repeat interviews with the same people over a long period could have been even better. She did not add that such studies are notoriously expensive.

What Sharma uncovered was not one simple decision to consult a complementary practitioner, but what she describes as 'chains of decisions taken over quite a long period'. The reasons her 30 interviewees first consulted echoed those of other studies: non-life threatening conditions, chronic conditions, and those where mainstream medicine had, in the eyes of the patient, been less than satisfactory. But the picture was more convoluted. People needed additional reasons to continue – or not – consulting a complementary practitioner, hence her evocative '*chains* of decisions'. She summarized these interlinked decisions by sketching three different types of user.

One she described as the 'earnest seeker'. This included those who had not yet found a cure, but intended to keep trying and those who were neither satisfied nor deterred. A young man declared, 'I have got very high standards for health'. He had used non-orthodox medicine for severe eczema for four years and had consulted spiritual healers, a reflexologist and an acupuncturist, and was having homeopathic treatment. 'Although he did not feel that his condition had improved very much as a result of all this effort, he appreciated certain features of the non-orthodox therapies he had used, especially the time given by the practitioners to diagnosis and their preparedness to discuss treatment with the patient', Sharma reported.

'Stable users' was her second type. These either regularly use some type of complementary therapy for a particular problem or use a single form of complementary medicine for most problems. One such had started to have homeopathy before the Second World War to deal with a severe ear infection. He and his wife had used homeopathy for most health problems ever since, even treating himself with it for routine ailments.

The third type Sharma labelled 'eclectic users' – those who used different forms of complementary medicine for all kinds of problems.

Sharma's use of qualitative research captured the multi-faceted nature of complementary medicine use, something that is far harder to pin down with fixed-choice questions typical of social surveys.

A small footnote to Sharma's work is worth reflecting on. Given that one of the features of complementary practice is held to be the quality of the relationship between patient and practitioner, and given that it is often characterized in terms of the length of consultation, it might be important to distinguish between the nature of the therapy and the context in which therapy is being offered. Complementary medicine is private medicine, i.e. a fee for a service, payable at the time. In Stoke-on-Trent, this was still cheaper than having private treatment via allopathic medicine. Research available at the time Sharma was working had already indicated that being able to 'buy' time was one reason for seeking private conventional medical care. One of her interviewees, educated and middle class certainly, but not at all wealthy, said:

> The attention you get is important. Going private (for an earlier skin condition) I did get a chat and an explanation. I was treated as an intelligent person. It is worth paying for that, although you should not have to. Jenny (the herbalist whom she was currently consulting about her child's

> allergy) does explain what she is doing and I feel more active in the treatment.

○

Christine Barry, Senior Research Fellow in Medical Anthropology at Brunel University, argues that surveys rely on an idea of human behaviour as explicable with reference to internal, presumably psychological processes. Social anthropologists explain behaviour with reference to social interaction with others and pay attention to the context in which the behaviour occurred. A social survey conducted separately from the occasion in which the behaviour in question arose loses these features. In any case, Barry points out what is very well known: there is often a demonstrable discrepancy between what people say they did and what they actually did. Rather than use this discrepancy to discredit what people say, she instead found that inspection of this very discrepancy can itself be revealing. Actually watching what happens, preferably over a long time, as a supplement to interviewing, is the best way of exposing and exploiting this discrepancy analytically, Barry suggests.

Integrated Medicine is a concept that has emerged over the last decade or so. It attempts to integrate complementary and orthodox medicine and, hopefully, offer the best of both. Integrated, or integrative in the USA, medicine courses and centres have sprung up all across the developed world. The Association for Integrative Medicine claims members from Germany, Pakistan, Mexico, Brazil, England, Romania, Canada, Hong Kong and India. An example of qualitative research offers an interesting insight into integrated medicine at work.

Christine Barry completed her PhD thesis in social anthropology, a discipline which has qualitative research at its very core in 2003. Entitled 'The body, health, and healing in alternative and integrated medicine: an ethnography of homeopathy in South London', it is a wide-ranging study of homeopathy. One part focused on the work of a general practitioner who was trained in homeopathy among other alternative therapies. Barry observed and audio-tape recorded 20 separate consultations at a South London clinic. She also audio-tape recorded interviews with the GP, Dr Deakin (not his real name), and those patients she observed. The hours of tapes of consultations and lengthy semi-structured interviews were transcribed for analysis verbatim. Overall, she observed 23 consultations with Dr Deakin and 23 with non-medical homeopaths, conducted 46 interviews and filled 24 notebooks with fieldnotes: a huge amount of detailed data on actual practice in everyday contexts.

Barry found that Dr Deakin gave many of his patients the choice of treatments by asking them 'Would you like antibiotics, herbal or homeopathy?'. Some of his patients liked the fact that he offered alternatives to standard medical treatments, while others were confused by having to choose. There was no time in the brief consultations for him to explain the rationale behind each therapy, leaving some patients muddled about the homeopathic remedies they had been prescribed; many ending ended up not taking them. Barry contrasted these observations with her time spent studying non-medical homeopaths, who had a completely different philosophy. They conducted longer consultations, free from the constraints of a publicly provided health service, focused purely on homeopathic treatment. Their clients came to engage fully with the alternative philosophy of homeopathic medicine, she reports.

Barry concluded that Dr Deakin is a hybrid – neither fully homeopath nor general practitioner. He is a better listener and more empathic than many orthodox doctors, but, she felt, applies the principles of homeopathy as if they were just another type of treatment rather than a total philosophy of healing. He was also restricted by working within a standard general practice unsympathetic to alternative approaches. For example, he was refused the longer consultation times that are required for more holistic treatments.

Barry's work raises an important question. Does integration itself change the very nature of the type of medicine that a doctor practices? As she says, '...what version of a therapy is being integrated, how the patients approach it, what constraints are limiting its application and how does it differ in different contexts?'. One conclusion is 'for certain patients, choosing alternative therapies is in itself a reaction against orthodox medicine, and so the offering of alternative medicine by an orthodox system becomes a paradox'.

As with so much research, Barry's work raises more questions than it answers; there is, as yet, no answer to her paradox. However, it does show how qualitative research, this time from anthropology, can offer new insights into how any form of medicine is practised.

○

The other type of qualitative work is Conversation Analysis (CA). Conversation Analysts have for some years studied consultations in conventional medicine and are now beginning to turn their attention to complementary medicine sessions.

This is very different from the work considered so far. To provide a minimum illustration of what is involved, some technical detail is necessary.

CA was developed in sociology and linguistics. Paul Drew, sociologist at York University and one of the leading lights of the discipline notes, it is 'an observational science: it does not require (subjective) interpretations to be made of what people mean, but instead is based on directly observable properties of data... properties (which) can be shown to have organized, patterned and systematic consequences for how the interaction proceeds'. Conversation Analysis focuses primarily, but not exclusively, on verbal patterns which recur as people take turns talking with one another. It also takes into account features such as direction of gaze or body orientation. CA depends on *naturally occurring* interactions, those that happen whether or not any research is under way. Such interactions are captured by audio or perhaps video. The researcher does not, therefore, ever have to be present. Recordings are then transcribed in great detail. Everything said is transcribed verbatim, and other features, such as intonation, ums and ers, volume, and pauses are annotated. A series of symbols has been developed (see Table 1) allowing as faithful a transcript as possible of taped conversation to be typed onto the page.

There are three features to grasp to get a handle on CA. Any utterance and many non-verbal actions – body movements, say – are part of the conversation under investigation. These sounds or actions make up a sequence in which each is influenced by the one that came before. These sequences appear to have regular patterns.

Above all, a fundamental feature of conversation is that people *take turns* even though they do interrupt each other from time to time. This concept of *turn design* is central to CA and is readily visible in transcribed conversation. For example:

01	Dr:	Hi Missis Mo:ff[et,
02	Pt:	[Good morning.
03	Dr:	Good mo:rning.
04	Dr:	How are you do:[ing

Table 1 Selected transcription symbols.

The relative timing of utterances	
Notation	Meaning
0.7	Intervals within or between turns shown as time in seconds
(.)	A discernable pause too short to time
[ ]	Overlaps between utterances, the point of the start of the overlap marked with a single left-hand bracket
=	No discernible interval between turns. Also indicates very rapid move from one unit in a turn to the next
Characteristics of speech delivery	
Various aspects of speech delivery are captured in these transcripts by punctuation symbols (which, therefore, are not used to mark conventional grammatical units) and other forms of notation, as follows:	
Notation	Meaning
.	Falling intonation
,	Continuing intonation
?	Rising inflection (not necessarily a question)
:	Stretching a sound, the number indicate the length of stretching
Italics	Marks the sound stress
CAPITALS	Emphatic utterance usually with raised pitch
( )	Unclear or uncertain utterances speech placed in parentheses

05	Pt:	[Fi:n]e,
06		(.)
07	Dr:	How are y[ou *fe[eling.*
08	Pt:	[Much [(better.)
09	Pt:	I feel good.
10		(.)

What is particularly nice about this example is the way that CA can illustrate how very similar phrases – *How are you doing* and

How are you feeling – uttered within seconds perform different actions. There is a slight difference in the construction of the turn (*doing* in the first, *feeling* in the second) and this reflects the contrasting actions which each performs. The first is an all-purpose general open-ended polite enquiry with which people frequently begin interactions. The second, though, is a medically focused enquiry inviting the patient to talk about whatever problem they have come to the doctor's surgery for. The patient distinguishes between the actions that each turn is designed to achieve. As Drew and colleagues observe: '(S)he responds to the former as a social enquiry with *Fine* (line 5) but to the latter biomedical enquiry with a form (*Much better. I feel good.* lines 8 and 9) which manifests her understanding that he is enquiring about her progress in coping with the condition about which she last consulted the doctor'. The point is that the patient's responses are connected to the designs of the preceding turns.

In contrast with other analytic approaches to audio-tape recorded conversations – be they interviews such as John Hughes did, or consultations such as Christine Barry analyzed – CA aims, say Drew's team, 'to identify and describe the specific interactional consequences which follow from given verbal practices'. CA typically entails very large collections of data: a study in the USA, for instance, included well over 300 consultations involving 19 physicians.

An example of the type of insight that CA can offer comes from the 2003 PhD work of John Chatwin, a student of Drew's, entitled 'Communication in Homoeopathic Therapeutic Encounters' and carried out at the University of York. Chatwin scrutinised 30 hours of homeopathic consultations recorded with 8 practitioners and 20 patients. He rarely found homeopaths directly telling a patient to stop taking an allopathic drug. 'It appears to be more common, when homoeopaths engage in talk about allopathic medicine, for categorical formulations to be attenuated and for comparatively subtle and sequentially extended approaches to be used', he comments. The homeopaths, Chatwin's transcription indicates, set up hints so that it was the patients, by and large, who suggested giving up the drugs themselves. Chatwin's extract looks like this.

1 Hom: So how many cortizol ((allopathic drug)) have you got
2 left
3 (0.5)
4 Pat: Oh: not many=
5 Hom: =not many=
6 Pat: =no not many about (.) erm [(??)
7 Hom: [And how long are you going
8 to be doing this no- nasal spray for
9 (0.5)
10 Pat: Erm (.) I see the consultant (.) I mean I could stop
11 it now if=
12 Hom: =Yea
13 Pat: really shall I stop it now
14 Hom: Yea IF IT'D WORKED I'D'VE SAID <u>NO</u> Keep going with it but
15 it's NOT WORKED .h [then I wouldn't bother
16 Pat: [No I don't think it has. . .

From this Chatwin concluded 'the homoeopath would clearly prefer the patient to undertake a particular course of action – in this case to stop using her prescription drugs – she approaches the issue in such a way as to let the patient be the one to bring this into the open'. This does not appear to be evidence of deliberate manipulation; rather it reflects the patient-orientated approach of the homeopath. The meticulous approach of CA has highlighted some of the subtleties of the homeopath–patient relationship. If such relationships are the key to alternative medicine's success, then maybe some hints to as to why lie in such studies.

There are signs that qualitative research into unorthodox medicine is being taken seriously. Christine Barry's post-doctoral work and three other qualitative projects are funded by the UK's Department of Health (following the House of Lords' Report) as part of a programme to strengthen research into complementary medicine. And debates about the standard of such studies are beginning to appear in the conventional medical literature. Helen Lambert and Christopher McKevitt, medical social anthropologists, writing in the *British Medical Journal*, warned of just picking up a qualitative research method and applying it without having a proper grasp of how to do so. They describe as misguided the habit of divorcing 'technique from the conceptual underpinnings'. In essence: the same rules apply to qualitative social science research as to medical research: understand the tools before you use them!

7

hints of understanding

It's time to stop dancing around the placebo effect and meet it head on. Clinical trials use placebo as a mark of failure; if something works no better than a sugar pill then it is not effective. Doctors embrace the placebo and cheerfully admit that it is an important part of their toolkit. Complementary and alternative practitioners say that their therapies work by encouraging the body to heal itself – a placebo effect – and critics argue that this means they do nothing. Nevertheless, research is beginning to suggest that the placebo effect may play a role in *every* act of treatment.

The word placebo comes from Latin, meaning 'I shall be pleasing'. The *Oxford English Dictionary* illustrates its use in medical terms thus: '...an epithet given to any medicine adapted more to please than benefit the patient'; 'It is probably a mere placebo, but there is every reason to please as well as cure our patients'. In other words, according to the *OED*, a placebo is not a useful treatment but a sop to a patient. These quotes, though, are from 1811 and 1888. Medical thinking has moved on a lot since then. In a paper published in the *American Journal of Pharmacy* in 1945 O.H. Perry Pepper described the placebo as 'the art of human care' and contrasted it with 'the science of medicine'. Complementary medicine researchers concur. Professor Paul Dieppe describes complementary therapies as being centred around the belief that 'caring for the patient' brings relief and is content that they may be largely based on a placebo effect.

Doctors have known about and exploited placebo for thousands of years. Hippocrates (c. 460–377 BC) clearly understood the power of the patient's mind. 'The patient, though conscious that his condition is perilous, may recover health simply through his contentment with the goodness of the physician', he said. Indeed, the Hippocratic Oath enshrines this understanding that it as important to care for the patients' general well-being as to treat them for their specific condition.

'... that warmth, sympathy, and understanding may outweigh the surgeon's knife or the chemist's drug'. Yet it is unethical, in the strictest sense, for a doctor to prescribe a sugar pill. And with good reason, as it is crucial that patients know what they are prescribed. Someone cannot give informed consent for treatment if they don't know what it is. Patients also need to know what drugs they are on in case they are taken ill and need to see an emergency doctor who doesn't have their notes to hand, or in case they consult another practitioner who needs to check for contra-indications.

In clinical trials placebo can have a huge economic impact: it is the standard by which newly discovered drugs are measured. One of the tests a drug has to pass on the way is whether it is more effective than a placebo in a clinical trial, or more normally in a series of trials that can include thousands of patients. In October 2000 the share price of UK-based biotech company Cantab Pharmaceuticals dropped 67% after its developmental drug to treat genital warts was found to be no better than placebo. A similar fate befell British Biotech when the first results from a programme of 10 randomized controlled trials cast doubt on its anti-cancer drug Marimastat. The *British Medical Journal* commented 'According to the industry, these huge falls illustrate how clinical trials are being closely tracked and the results used by financial analysts to make and destroy fortunes overnight. Analysts, they say, are now as likely to be found reading the *BMJ* and *The Lancet*, for clues about the progress of trials, as they are the *Financial Times*'. In financial terms, 'no better than placebo' can hurt.

It is inevitable, important even, that many drugs fall by the wayside during the development process. Companies know and accept this and normally have a portfolio of options under development to compensate. They are explicit about the risks, often including disclaimers alongside their development portfolio such as this one from the multinational giant GSK: 'Owing to the nature of the drug development process, it is not unusual for some compounds, especially those in early stages of investigation, to be terminated as they progress through development'.

For big pharmaceutical companies and their investors, placebo is firmly a negative. A drug that does not perform better than a placebo is a drug that will not make money. However, just because a drug does no better than a placebo does not mean that it does nothing. The simple act of administering a placebo does produce an effect.

Working out exactly what is going on is not simple, particularly when it is possible to produce a placebo effect without actually giving a placebo. A randomized clinical trial published in *Clinical Trials Meta-analysis* in 1994 found that if patients were told what to expect when given a drug they duly showed a bigger response to it. The act of giving the patient more information enhanced the effect of the drug. That is, the information alone elicited a placebo effect, yet no placebo was administered.

Some researchers refer to this placebo-effect-without-a-placebo as a 'context effect'. Context effects are the impacts of almost everything else other than the drug, surgery or whatever in question. They take into consideration all the factors within a consultation: the physical environment, the relationship between the patient and the practitioner, the amount of discussion between the two and the expectations of both. They can include everything from the moment a patient first becomes ill to the point at which they finish the treatment and stop seeing the doctor. Suddenly, what goes on between doctor and patient starts to look very complicated indeed.

Attempting to work out what is context effect and what is treatment is a significant challenge. A study that appeared in *The Lancet* in March 2001 identified 25 clinical trials that included an element of context effect. Author Zelda Di Blasi and her colleagues split these effects into two categories, cognitive care and emotional care. The former aims to improve the patient's understanding and expectations of the treatment; the latter seeks to make the consultation a more relaxed and less fearful experience. The researchers found the trials to vary considerably and did not draw any firm conclusions. The one consistent observation they made was that 'physicians who adopt a warm, friendly and reassuring manner are more effective than those who keep consultations formal and do not offer reassurance'.

What is treatment and what is context becomes even more blurred for complementary therapies. A common theme from homeopathy to shiatsu is providing a good environment for the patient, one that is welcoming, encouraging and unhurried. A typical therapy room has soft lighting, comfortable furnishings and perhaps a candle burning and soothing music. Therapists take time to make their rooms as pleasant to be in as possible. Complementary practitioners tend not to split environment from treatment. They see the place where someone receives their therapy as part of that therapy. In other words, practitioners deliberately seek out context effects. Contrast this with the orthodox medical view that a drug will be efficacious whether it is administered in a plush suite or on a battlefield. That is not to say allopathic doctors do not care about where they practice, just that environment is not as deeply linked to treatment in the same way.

○

A good argument for seeking a positive placebo effect is that it also has a dark side. Derived from the Latin meaning 'to harm', nocebo is placebo's evil twin.

This idea has probably been around for as long as we have. Concepts such as being 'scared to death' or 'putting a curse on' an enemy are well established in all cultures. Haitians believe a voodoo curse can kill – victims are not poisoned or injured, just told they are going to die. Sicilians believe that a hex can cause headaches and versions of 'evil eye' superstitions are found right across the Mediterranean, Middle East and Asia. There is far less research into nocebo than placebo, but what has been done is arresting.

An oft reported study was conducted on 57 Japanese schoolboys to determine their response to allergens such as pollen and nuts. The boys filled out questionnaires relating to their past experiences with rashes and skin problems. The researchers then selected those boys that had reported sensitivity to the lacquer tree, which, like poison ivy, gives some people a rash. The scientists then blindfolded these boys and told them that one arm was about to be brushed with the leaves of the lacquer tree and the other with chestnut tree leaves. The researchers then brushed chestnut leaves on the arm the boys thought would receive lacquer tree leaves and vice versa. Almost immediately, most of the arms that had expected lacquer tree foliage developed red, irritated rashes. The other arms remained unblemished – despite having touched the aggravating plant.

Another famous case was part of the Framingham Heart Study. This long-term study of cardiovascular disease was set up in 1948 and funded by the US National Heart Institute, now the National Heart, Lung and Blood Institute. Over 5000 inhabitants of the small town of Framingham in Massachusetts were recruited in the first wave; 23 years later another 5000 signed up. As the *Journal of the American Medical Association* reported in 1996, the project revealed that women who believed they were prone to heart disease were four times as likely to die from cardiac problems as women who had exactly the same risk factors but did not believe they were in any additional danger. The conclusion the researchers reached was that the mindset of the first group contributed to their ill health. They thought themselves ill.

Designing experiments to study nocebo is an ethical morass – to do so would be to set out to cause harm. In the case of the Japanese schoolboys the harm was small and so considered justifiable. The nocebo effect from the Framingham Heart Study emerged from data analysed retrospectively; clearly it would have been impossible to deliberately try to induce a life-threatening condition by suggestion.

Then there is the additional ethical question of informed consent. Part of informed consent is telling the participants of any known side-effects of the what they may or may not be getting. The implication from what we know about nocebo is that if you tell someone that they might experience tiredness and nausea, say, then they are more likely to report feeling tired and nauseous. This is particularly problematic when a new treatment is being compared with a well-established one. What should each group be told? Is it ethical to tell someone that they might get a certain side-effect when it has not been reported for that drug but instead for the experimental medication? Does that then produce an artificial set of side-effects that continue to be ascribed to the new drug? No one knows.

○

Much of the evidence for placebo, nocebo and context effects comes from clinical trials, which, by their very nature, do not offer explanations for how the effects appear. Implied in all of them is the fact that thoughts or state of mind can have an effect on health, one of the tenets of all complementary medicines. Much of modern medical research has its origins in the mechanistic model of Cartesian duality – splitting the mind from the body and treating it as a machine. The truth is not that clear-cut. Doctors certainly recognize the importance of a patient's state of mind and the literature is full of papers reporting it and discussing the implications. A major contribution comes from a new branch of science that is finding evidence of direct links between mind and body.

By the end of 1970s there were enough immunologists, psychologists and neurologists with similar interests working in similar fields for the discipline of psychoneuroimmunology to be born. The phrase was coined by Robert Ader, who was responsible for some of the seminal work in the field now best described as the study of how state of mind influences the immune system and vice versa. Psychoneuroimmunology is now an established discipline with its own journal and a professional society that has met annually since 1993.

Broadly, our immune system has two roles: to fight off infection and to identify and destroy rogue cells before they can turn into cancers. It is often likened to a defending army charged with both repelling boarders and putting down insurrections. Like an army it has different weapons that can be applied in different ways. Similarly, it has an elaborate control system designed to keep it in check, because an uncontrolled immune system is just as dangerous as an uncontrolled army. Fast-moving mobile scouts spot the enemy; slower but more powerful heavy artillery can be called to the trouble spot.

When the immune system's scouts detect an invading bacterium or virus, they send out signals summoning the heavy squad to kill the invaders. These signals are chemicals called cytokines. Amongst other things, cytokines increase the flow of blood to a damaged or infected area and make blood vessel walls leaky so that white blood cells, the heavy artillery, can get to the trouble spot. Having made a problem area warm, red, swollen and painful, cytokines attract reinforcement white blood cells as long as the infection or damage persists. The cytokines involved in the inflammation response – pro-inflammatory cytokines – are central to current research in psychoneuroimmunology.

Over-production of one pro-inflammatory cytokine, Interleukin-6, is associated with many chronic degenerative illnesses, including cardiovascular disease, osteoporosis, arthritis, type 2 diabetes, cancers, gum disease and general frailty and decline. A study published in 2002 in the *Proceedings of the National Academy of Sciences* demonstrated how chronic stress effects levels of Interleukin-6. Janice Kiecolt-Glaser and co-workers identified 119 people who were caring for partners with dementia and compared them to 106 similar people without caring responsibilities. These were sufficient numbers for a valid conclusion. The carers produced on average four times as much Interleukin-6 as the non-carers. Surprisingly, this level of overproduction continued for some years after their spouses died. The team argued that these data offered a mechanism by which living with stress could translate directly into chronic diseases.

This study is part of a larger body of research that connects social situation with well-being. One rather elegant paper showed a strong link between having a good network of close friends and relations and the ability to fight off colds. Another found a correlation between high levels of stress hormones in the first year of marriage and the likelihood of divorce. All told, psychoneuroimmunological research has demonstrated that exercise, stress, depression, sleep, social isolation, accident or bereavement can have effects on health issues from susceptibility to infection to cancer. And the traffic between mind and body is two-way. Glitches within the immune system can feed back to the brain and precipitate changes in mood and perception.

Alongside psychoneuroimmunology are other studies that are beginning to reveal ways in which specific placebo effects might work. Some of the most compelling have been carried out on patients with Parkinson's Disease, the degenerative neurological disorder that affects walking, writing, speech and so on. Typically there are three main symptoms: tremors, muscle stiffness and slowness of movement. People with Parkinson's can also get depressed and tired, and find it difficult to communicate. It is a cruel condition that slowly and relentlessly gets worse.

The disease is caused by the gradual death of the cells in the brain that produce the essential chemical messenger dopamine. There is no cure, but taking synthetic forms of dopamine can help and there has been some success with transplants of dopamine-producing cells into the brain.

Doctors working with Parkinson's patients have noticed that in some stressful circumstances, such as the need to escape from a fire, normally immobile patients can become briefly mobile. Two suggestions have been put forward for how this 'kinesia paradoxica' might work. The body might have an emergency escape system that does not rely on dopamine, or stress may cause a burst of dopamine in the brain. Thought, that is to say, can have a direct metabolic effect.

Alongside this phenomenon there have been a number of studies that indicate that the placebo effect is particularly strong in Parkinson's patients. It is often the case, for example, that those in the placebo groups of clinical trials show a significant improvement.

A series of experiments from the Pacific Parkinson's Research Center at the University of British Columbia has suggested a mechanism for these observations. They have also provided hard evidence that administering a placebo can cause physiological changes. The researchers used a type of brain scanning called Positron Emission Tomography, or PET. This technique measures the amounts and locations of chemicals in the brain – in this case, chemicals associated with dopamine release. One group of subjects was given a common Parkinson's drug and the other an injection of saline. Both groups produced a surge of dopamine in response. Those given the drug had a slightly bigger surge, but the differences were relatively small.

The researchers propose that this placebo effect works through the brain's reward mechanism. This general brain circuit unleashes dopamine in response to or anticipation of activities such as eating, having sex, or some other behaviour producing a feeling of well-being. In the main it is a valuable encouragement to animals and people to do things that are good for them; if there was no reward for eating or copulating then the species would rapidly die out. The circuit can be hijacked though. Addictive drugs like alcohol, heroin and cocaine stimulate dopamine release.

The reward mechanism appears to be tickled by the anticipation of receiving the treatments, not receiving them. The Parkinson's patient anticipates that the placebo injection will have an effect, and this activates the reward mechanism which produces a surge of dopamine. It could also explain kinesia paradoxica. The anticipation of getting away from a source of stress may activate the production of the dopamine that the brain needs to tell the muscles to work.

This neat piece of research offers a clear biochemical explanation for one placebo effect. But as one of the researchers pointed out in the journal *Science* in August 2001, these results 'suggest that in some patients, most of the benefit obtained from an active drug might derive from a placebo effect'.

The dopamine reward pathway is not necessarily a general mechanism for all placebo effects and it is probable that different conditions may have a different susceptibility to placebo. There is almost certainly a variation in the way different people respond to placebo as well. Nonetheless this research demonstrates two things: that the placebo effect can be real, measurable and physiological and that it can make a significant contribution to the improvements that patients show after treatment. Likewise psychoneuroimmunology does not provide a universal explanation for why treating a patient well makes them better. It merely provides evidence of a link between social condition and health. Both these areas suggest some important questions for future studies to tackle. Not least, do complementary therapies elicit physiological effects similar to those seen in the experiments above?

○

George Lewith is Senior Research Fellow at the University of Southampton and Visiting Professor at the University of Westminster. He is a medical doctor and an experienced acupuncturist, and has been researching alternative medicines for years. He describes one of the key elements of a complementary consultation as 'listening with an intent to heal'. This, he argues, is fundamentally different from just providing a sympathetic ear.

Complementary consultations range from 30 to 90 minutes, compared to the seven to 20 minutes or so that a general practitioner offers. Lewith, though, contests the idea that if every doctor could spend an hour with each patient then they too might make better headway in seemingly intractable situations. This, he argues, is to misunderstand the nature of the therapist's role. If a patient has a chronic condition that orthodox medicine cannot treat then a conventionally trained doctor might listen for three-quarters of an hour feeling impotent in the knowledge that there is nothing by way of treatment to offer. By contrast, says Lewith, an acupuncturist might not feel they can cure the patient, but may well believe they can make them feel better. So they listen with the hope and expectation that they can do something. While the patient tells their story, they will be thinking 'I wonder if spleen 6 would work here' or 'There has been an invasion of damp, what can I do about it?'. It is an active, participatory process, different, Lewith contends, from a passive orthodox doctor spending time just for the sake of it.

It is a hard idea to test, but Lewith is running a study to see if he and his co-workers can identify an effect from 'listening with intent'. The team is recruiting women with undiagnosed severe pelvic pain. This chronic condition can ruin lives. These women have seen specialist after specialist yet are still suffering. All attend a pain clinic and have had considerable amounts of specialist medical intervention. This, says Lewith, suggests that it is not just a matter of time with a doctor that counts. After recruitment these women will be given a Traditional Chinese Medicine consultation and diagnosis. Crucially, the women will be offered an explanation of their condition in Traditional Chinese Medical terms. This diagnosis is unlikely to have any grounding in modern medical science, but the point is to investigate whether providing a model for understanding can influence the condition.

Lewith also plans a secondary study to ask further questions about what, as he describes it, 'sets people up for success'. Is it belief, depression or state of mind? Is it possible to separate the process of diagnosis from that of treatment? He is trying to find out what, if anything, is special about the way in which a complementary practitioner practises compared to an orthodox doctor.

○

Our growing understanding of placebo, nocebo and context effects is challenging to some complementary practitioners. It implies that what they achieve has nothing to with the treatment they offer and everything to do with the way they do it. More positively, it hints that context effects need to be carefully monitored, nurtured and controlled to reach their maximum efficacy. Throwing someone into a cold, stark room and sticking needles into them without any preamble is far more likely to elicit nocebo than any benefit. Medical research may never find any other effect than placebo for some complementary and alternative medicines. But if the placebos they produce are bigger, more powerful and more effective than any others then these too could be considered powerful healing methods that medicine might wish to harness reliably, rather than dismiss.

> The nocebo and placebo phenomena remind us of the complexity of healing and the importance of the art of medicine.
>
> Malcolm P. Rogers MD, 2003

8

a new model

The rise of testing of complementary therapies has mirrored but lagged behind the growth in their popularity over the past two decades. Clinical trials of various types have probed orthodox treatments over the past 50 years or so but there is a far shorter tradition of using them in alternative medicine. Then there are all the added subtleties of complementary therapies: outcome, context, therapist–patient relationship and so on. The result is a relatively small group of people familiar enough with all the threads to draw them together coherently.

One such is Andrew Vickers of New York's Memorial Sloan-Kettering Cancer Centre. In 1996 he published a paper in the *Journal of Alternative and Complementary Medicine* entitled 'Methodological Issues in Complementary and Alternative Medicine Research: a personal reflection on 10 years of debate in the UK'.

Vickers outlined his view on the shortcomings of conventional medical research techniques in the study of alternative therapies. He did not call for a whole new set of tools, instead arguing that randomized controlled trials can be adapted. He summarized his thoughts in a table that matches research questions with study design.

The question that a researcher wants to answer is the key to all good scientific endeavour, whether in anthropology, physics, genetics or acupuncture. A clear, well-formed query will lead more easily to useful results. Much of the challenge of any project is formulating that initial question. (Incidentally, good qualitative data on complementary medicine should help define good questions for quantitative research.)

Vickers took acupuncture as a model therapy and tabulated 10 types of research question that might be asked about it and the type of study best suited to answering them. For example:

Question: Is acupuncture an effective treatment in practice?
Research Design: Audits of practice with long-term follow-up using validated outcome measures; comparative cohort studies; pragmatic research: randomized trials comparing acupuncture as a package of treatment to standard care.

Question: Is acupuncture a placebo?
Research Design: Fastidious, randomized, placebo-controlled, blinded trials.

The rest of the table covers other questions such as what type of conditions acupuncturists treat or what patients experience. Each draws on a different set of research methods. It's a lucid illustration of how different methods can offer different answers and why using more than one method is essential to getting a complete picture.

One of the leading proponents of properly controlled clinical trials of complementary medicine is Edzard Ernst of the Universities of Exeter and Plymouth. He trained and practised as a doctor in Germany and studied a raft of complementary therapies including acupuncture, autogenic training, herbalism, homoeopathy, massage therapy and spinal manipulation For the past 10 years, Ernst has been conducting research into all forms of complementary medicine and has published prolifically in the field. He and his colleagues have compiled the booklet *The Evidence So Far*, mentioned previously; it is a synopsis of clinical trials and studies that the Department of Complementary Medicine has either done or reviewed. A broad document, *The Evidence So Far* covers many different areas, from osteopathy, chiropractic and acupuncture to homeopathy.

The key questions in complementary medicine according to Ernst are those of medicine as a whole, namely: 'Is it safe?' and 'Does it help anybody?'. Ernst is evangelical about the power of rigorous research to answer these questions and is adamant that good research comes from collaboration. It is no good attempting a trial in acupuncture, he says, unless there is an acupuncturist, an expert in trial design and a statistician on board from the outset – it is rare to find all these skills within one individual. However, to design such a trial is expensive; it costs money just to convene the investigators before any patient recruitment even starts. Ernst reckons that one reason why so many complementary studies are weak is lack of money. A major consideration for the future of this type of research is funding. Some money comes from the manufacturers of herbal medicines, aromatherapy oils and flower remedies, but it is unreliable and paltry next to the huge sums spent by big pharmaceuticals companies.

His question 'Is it safe?' is only partly answered by randomized controlled trials. They pick up obvious hazards, but the numbers of participants are too small to detect rare side-effects. It is better to do retrospective epidemiology by examining patient records. Again, Ernst emphasizes the need for rigour and teamwork.

Ernst has big concerns about the placebo effect. As a clinician he wants his patients to get better and concedes that *how* that happens takes second place. But as a researcher he finds it very difficult to sanction a treatment that appears to have no intrinsic effect. It is regressive – a huge step backwards into the dark ages of medicine – because it doesn't lead anywhere.

That said, Ernst does not disregard the placebo effect; he believes it should be studied in greater detail. Why a spiritual healer elicits a huge placebo effect where a doctor does not is a fascinating research conundrum. The phenomenon is important to understand, he says, as all practitioners want to exploit it.

He has mixed views on qualitative research. The good stuff, he recognizes, can answer important questions about the meaning of illness or suffering. But it has little to say about efficacy. Worse, he cautions, there has been a proliferation of poor qualitative research, often questionnaire-based. This can be dressed up as proving something that it is incapable of doing, he warns. He calls it 'politically correct' research that does not provide hard answers but does not upset anybody.

Professor Ernst's views have earned him opprobrium from many quarters, and he has the letters to prove it. Nevertheless he stands by his insistence that scrupulously controlled, properly designed clinical trials are powerful tools in the study of complementary medicine. This scientific approach has won over a number of medical consultants, he argues, who now approach the subject in a much more positive light.

Medical researchers and complementary therapists don't talk to each other enough, agrees Professor Stephen P. Myers, Director of the Australian Centre for Complementary Medicine Education and Research, a joint venture of the University of Queensland and Southern Cross University. Myers is curious about the influence of consultation and individualization on the effectiveness of homeopathy. Clinical researchers often design and conduct experiments into homeopathy, he points out, without ensuring that they are actually assessing what homeopaths really do. In an attempt to overcome this problem, Myers brought together trial experts and homeopaths for a brainstorming seminar. The study he is now overseeing is a direct result of this collaboration and is being undertaken by Don Baker one of his PhD students.

They have put together a set of osteoarthritis patients, confirmed by detailed diagnosis including X-rays of their deteriorating joints, who are prepared to take part in a placebo-controlled trial. To date they have recruited 115 of the 135 patients they seek to enrol. The patients are divided randomly into two groups; one has a consultation from a homeopath, one does not. These groups are then further subdivided. One arm of the no-homeopath group receives a complex of homeopathic remedies designed to treat osteoarthritis; the other arm receives a placebo. The consultation group is divided into three arms. The first receives the same homeopathic complex, the second a placebo and the third group an individualized homeopathic remedy.

This five-armed trial is designed to answer a number of questions. First: what is the effect of consultation on the efficacy of the homeopathic complex for osteoarthritis? This will show up in any differences between the groups that receive the complex and placebo, the only difference being that one set will have had a consultation and one will not. Second: what is the effect of individualizing homeopathic treatment? Any differences between the fifth group and the others will suggest that customizing impacts efficacy.

This is an elaborate exercise, designed by experts in clinical trials and homeopaths from the outset; it is an *experimental* experiment. It might not be perfect, but it should point the way to a reasonable next experiment. It is good science, in other words.

The idea that well-designed clinical trials can determine the efficacy of complementary medicine is echoed by Richard Nahin, Senior Advisor for Scientific Coordination and Outreach at the National Centre for Complementary and Alternative Medicine, part of the US Government-funded National Institutes for Health. He says 'the same techniques and methods used to study conventional biomedicine are applicable to complementary medicine'. Although he does point out that 'there may be some challenges in studying complementary medicine, such as randomization, blinding, and quality control, these challenges are not unique to complementary medicine'.

Nahin also recognizes the need for outcome measures to be relevant to what the trial is trying to do. In his view, 'If the purpose of the trial is to establish clinical efficacy, then state of the art measures of clinical efficacy specific to the disease or condition must be used. If the purpose is to assess changes in quality of life, accepted measures of quality of life must be used'. His position on the placebo effect, however, appears to be somewhat different from that of Ernst. Speaking on behalf the of NIH he said the organization as a whole is interested in 'interventions that can affect patient expectations and clinical outcomes regardless of their mechanisms of action.... In particular, we are interested in therapies that are not directly biologically active but that, nevertheless, produce changes in health by biological means through the mind–brain–body connection'.

Another approach is typified by the work of Aslak Steinsbekk of the Department of Public Health and General Practice at the Norwegian University of Science and Technology (NTNU) in Trondheim, Norway. He is a homeopath with a degree in sociology and so, perhaps inevitably, has a somewhat different viewpoint. In January 2004 Steinsbekk put forward a suggestion to the World Health Organization expert group on homeopathy. This described a strategy for research development for complementary therapies and other therapies widely in use but with a lack of research. This strategy is used by the National Research Centre on Complementary and Alternative Medicine, University of Tromsø, Norway, and can be compared to the development of pharmaceuticals that are to become licensed and even reimbursed.

The first stage in pharmaceutical development is the identification of a chemical that appears to have some sort of biological activity. Further experiments explore its mechanism, confirm its potential and lead to tests in animals. If these prove favorable the next stage is preliminary safety tests in humans. After this comes a series of steadily larger clinical trials to prove efficacy and effectiveness and to work out exactly how the drug should be used. Only then can it be licensed and released.

The path of a treatment that is in use on patients by the time researchers turn their attention to it, like complementary therapy, runs in reverse. The initial set of studies is to first describe the use; then they move on to safety before they determine the effectiveness of the therapy as it is given in everyday practice. If this turns out to be beneficial for the patients, then the efficacy of the specific components is investigated before finally moving on to the mechanism of action. The starting point for research into the effectiveness of pharmaceutical drugs is that they are an explanation looking for an application; complementary therapies are an application looking for an explanation.

When putting together a clinical trial, Steinsbekk's begins with the patient's perspective. Too many studies are done, he argues, to answer the questions that a researcher has about a therapy rather than what patients want from it. Results from studies so designed have no bearing on how patients use a therapy in real life. A trial might show that a particular homeopathic remedy has no effect on a heart condition, but if a patient wouldn't seek homeopathy for that complaint then so what? It's like spending time discovering that cotton wool makes a poor road surface – scientifically valid, but practically useless.

Also important, says Steinsbekk, is how patients are recruited into trials. For studies of orthodox medicine, subjects are often selected from specialist clinics. This doesn't work for complementary therapy, he argues.

Patients in a specialist clinic have usually been referred by other doctors. The first line of treatment probably hasn't worked for them and their condition is a little more tricky. This is very different from the group of patients that a complementary therapist might have on their books. These are likely a far more diverse group with an equally diverse set of symptoms and conditions. As a result, complementary therapists have to develop expertise in dealing with a wider set of problems. Clinic doctors and nurses, by contrast, are specialists.

For a study looking at the effect of a complementary therapy on diabetes, say, it would be only natural to recruit patients in the same way as for a drug study – from a clinic. The complementary therapists in such a trial would face a set of people with intractable diabetes rather than their normal cross-section of patients. They will be able to offer something from their repertoire, but it is not their focus. It is like asking a domestic central heating engineer to repair an industrial boiler. The engineer would make a good attempt at it, but probably wouldn't do as well as an industrial boiler expert.

Thanks to this design flaw, says Steinsbekk, complementary therapies are not tested for what patients use them for. They are tested instead for the way patients use doctors, and that can be very different, as some of his data suggests.

Steinsbekk has carried out surveys of patients' wants and expectations of complementary medicine – why, for instance, do parents take their children to homeopaths? In Norway, one in four of those visiting homeopaths are children under 10. The commonest complaints are skin conditions and upper respiratory tract infections – coughs, sneezes, earache and the like. Steinsbekk discovered an interesting trend amongst parents. If they were unsure what was wrong with their child they took them to an orthodox doctor. If they knew, or thought they knew, what the problem was, they took them to a homeopath. Clearly parents were not using orthodox doctors and homeopaths in the same way, but were consulting the latter 'as a supplement', says Steinsbekk. He also found that parents seem to choose homeopathy on the basis of recommendations from friends and relatives. Some took their children along despite being sceptical. They did not even need to believe in the treatment, it appeared, being more swayed by others' reports than their own convictions.

An important consideration for Steinsbekk is whether people are wasting their time and money going to see a complementary practitioner. It is only possible to assess this, says Steinsbekk, if they are considered in the context in which the patient pays for them – another reason to ensure that any research is conducted with a view to what the patient, rather than the researcher, wants.

But there are those who feel that biomedical research threatens the very existence of some complementary therapies and could itself harm patients. Paul Dieppe of the University of Bristol outlines a worst-case scenario. A series of clinical trials of, say, a homeopathic remedy, consistently turns up no evidence that it is better than placebo. As a result it is banned. This has a knock-on effect within the complementary medicine community and, for example, a herbalist becomes alarmed that their treatments are also under threat. Consequently, the herbalist adds an ingredient, perhaps an unlicensed use of a steroid, into the remedy that makes it appear to work but dangerous. This is not a completely fanciful idea. There are well-documented cases of apparently very effective herbal remedies containing high levels of strong steroids. Therefore, Dieppe argues, the focus of research into complementary therapies should be on safety, not efficacy. Steps should be taken to ensure that complementary therapists do not do damage by, for example, encouraging cancer patients to forsake conventional treatments.

This does not mean that Dieppe believes research should be abandoned. Far from it – his team is running several studies exploring outcomes and the context and placebo effects. His real concern is that 'CAM practitioners are allowing biomedical researchers to prove that what they do is a waste of time, even when it is obviously valuable'. In doing so, complementary medicine risks being consumed by the medical profession and losing its identity and unique mode of care. It is indisputable that many people, particularly with chronic conditions, experience great benefit from complementary therapies. No amount of research is going to change those individual experiences, argues Dieppe: 'leave them to care and allow people to be cared for'.

○

So what do trials of complementary therapies look like at the moment? Nurse Jenny Gordon is a Research Training Fellow at Napier University in Edinburgh, Scotland. She has an interest in childhood constipation, a condition she became aware of while working in a surgical ward in the late 1990s. Children with intractable constipation would return to the clinic regularly in great distress and be referred to a senior doctor who would treat them according to their expertise. A physician would prescribe laxatives, or a surgeon might operate to remove impacted faeces. Gordon was troubled that the children were receiving crisis management rather than help to overcome the problem. Constipation might appear quite straightforward, but it can be a highly complex problem, which is what makes it very difficult to treat.

A bit of digging convinced her that childhood constipation was generally poorly managed and little researched. There were very few clinical trials into how it might be tackled and scant evidence-based medicine for its treatment. Indeed, doctors couldn't even agree on a clear definition of what symptoms a child with chronic constipation shows. Concerned, she and her colleagues began to consider alternative, less traumatic and more systemic ways of treating constipated children.

As well as being a nurse Jenny Gordon is also a trained reflexologist and was aware that many patients noticed increased urine production and bowel movements after a reflexology session. She decided to see if she could turn this side-effect into a treatment. She ran a small pilot study of 70 constipated children. Those who received reflexology seemed to fare dramatically better than those that did not. However, it was a pragmatic study with poor controls and no randomization – children were given the treatment according to which ward they were on – so Gordon and her colleagues could draw no hard conclusions. The only thing they could say was that it warranted further investigation.

One of the puzzling things Gordon noticed was that many constipated children did not take their prescribed medications. This angered the doctors, who felt their time was being wasted: why would children fail to take a medicine that could ease their distress? The constipation had a huge impact on the children's lives: they couldn't stay overnight with friends, go away to camp or take part in many normal childhood activities. From her experience as a paediatric nurse, Gordon knew that treating children is frequently complicated by the attitudes and beliefs of the rest of the family. Paediatric nurses, like complementary therapists, are used to working in ways that take into consideration all aspects of children's lives, social, emotional, spiritual and so on. She suspected the children were not complying with treatment partly because of the complexities of family life. She had heard, for example, of a grandmother advising against taking a laxative because it 'makes your bowel lazy'.

Gordon decided that before she could start a full-scale randomized controlled trial she needed to find out more about the child's point of view. Why were some resistant to taking medicine? What other factors might be influencing the way they responded? She designed a qualitative study based on interviews and focus groups and put it to the hospital ethics committee. They rejected it. She reworked it and tried again.

They rejected it again. No matter what she did she could not get ethics committee approval for this preliminary qualitative research. The ethics committee felt that there were special circumstances: they were particularly concerned about the implications of interviewing children about their toilet habits. Gordon, an experienced paediatric nurse, could not allay their fears.

By this stage time was getting short and the team decided to go ahead with a randomized controlled trial for reflexology and childhood constipation. They drew up a proposal, put it to the ethics committee who approved it without a murmur.

The study is dividing children with childhood constipation at random into three groups. One will continue with normal treatment, one will receive foot massage but not reflexology and one will receive proper, simple reflexology. The aspect that Gordon is most anxious about is that the massage and reflexology will be administered by the children's parents, not a qualified therapist. The researchers will teach the parents how to do a simple reflexology session, and ask them to follow a particular pattern of sessions at home. The investigators have no control over whether the parents administer the procedure correctly and at the right time – or at all, in fact. However, Gordon hopes the trial will help the parents and children take responsibility for treatment and, pragmatically, it means that her team can see far more children. Parents will receive support but won't need to make a weekly journey to the clinic.

Despite this concern, Gordon believes the results will be significant as the trial is double blind and properly randomized. She is adamant that randomized controlled trials are central to the investigation of complementary therapies, but they may not look quite the same as mainstream ones. The point, she says, is to recognize these differences, understand them and stand up and explain very clearly exactly what each element of the trial is for and why it is being done in that way.

At the time of writing Gordon and her colleagues have recruited half the 180 children they want and results are starting to come in. Since Gordon is blind to which group is which she couldn't tell me what the preliminary results were even if she had wanted to. She is also gathering some qualitative data about the children's situations – not as much as she would have liked, but enough, she hopes, to set the results of the trial in a wider context.

That this study is attempting to measure a more natural situation – parents giving their children reflexology at home – addresses one of the common concerns about clinical trials. They measure, many contend, what would go on in an ideal situation, not what happens out in the real world. This is true for all types of medicine, but once again it can have particular relevance to complementary therapies. An illustration of this was conducted by George Lewith (he of the previously mentioned pelvic pain trial).

Lewith and his colleagues wanted to compare the placebo effect in a clinical trial with what might occur in a more realistic situation, such as a doctor's surgery. They divided people with chronic fatigue at random into two groups and gave one a placebo and the other a supplement designed for chronic fatigue. They then asked the patients which group they suspected they were in. Ninety per cent replied: the placebo group.

The researchers then strove to match more closely the real-world situation, with an 'open label' trial. The supplement was given in its original packaging. All the participants in this round showed significant improvements even though the supplement was exactly the same as the one they got before in an unmarked bottle. In short, had this supplement been tested in a randomized controlled trial it would have come out negative. Yet a patient prescribed it or buying it from their own money might well experience a benefit.

Because of these layers of influence, says Lewith, he no longer conducts quantitative trials in isolation: every one he proposes includes qualitative data collection exploring the range of influences on why patients did or did not get better.

Studies that include an element of qualitative research are becoming more common in orthodox as well as complementary medicines. The *meta*Register of Controlled Trials currently logs 36 such hybrid studies, looking at conditions as diverse as knee pain, epilepsy and hypertension. This is a comparatively new trend, but in time all medical research may be a blend of qualitative and quantitative – who knows?

9

the evidence leads

'A paradigm shift is underway in health care. It will change medical practice in the years ahead', wrote John M. Eisenberg MD, Director of the Agency for Healthcare Research and Quality, the health services research arm of the US Government, in January 2001. Eisenberg was discussing – in the publication *Expert Voices* from the US National Institute for Health Care Management Research and Educational Foundation – the move away from what he called 'hand-me-down' medicine to evidence-based medicine. He did not use the phrase 'paradigm shift' lightly, taking pains to reference its original use by the philosopher of science Thomas Kuhn apropos a revolution in scientific thought.

Evidence-based medicine was first named in 1992 by researchers at McMaster University in Canada. It is a process by which doctors seek and apply the best treatment for their patients. At first glance evidence-based medicine sounds like little more than formalized common sense. Every doctor wants to make the right decisions for their patients; surely only a negligent practitioner would knowingly do otherwise?

A much-quoted *British Medical Journal* editorial published in January 1996 outlined some of the key differences between evidence-based medicine and doctors simply applying the best of their knowledge. Evidence-based medicine is, said the article, 'the conscientious, explicit, and judicious use of current best evidence in making decisions about the care of individual patients'. That is, not just the evidence an individual doctor has acquired during a lifetime's practice – one person's experience – but the best evidence available from every source, every study, every clinical trial. This does not mean that doctors should disregard their clinical experience to become medical automatons. As the article goes on, the discipline: '...integrates the best external evidence with individual clinical expertise and patients' choice, it cannot result in slavish, cookbook approaches to individual patient care'. Clinicians are being exhorted to seek out the best possible evidence gathered by the medical world at large and then apply it to the patients they have before them.

Alongside evidence-based medicine is another important movement, evidence-based health care. This is about making the best decisions for groups of patients: whether to offer a new drug for a particular condition, open a new clinic in a particular region or remove a certain treatment from those available under an insurance scheme, for instance. As evidence-based medicine becomes the dominant way to treat individuals, so evidence-based health care is becoming the preferred way of making large-scale decisions – in other words, how much money is spent and on what.

'Evidence' refers to information that has been obtained by some form of medical research and then published in a medical journal. All the clinical trials discussed in this book will, if and when they are published, form part of this pantheon. The amount of evidence available is enormous and growing. The US National Library of Medicines runs a free electronic database, PubMed, that anyone can access online. This contains over 14 million biomedical publications stretching back 50 years, any one of which might contain information relevant to a patient sitting in surgery.

So let's imagine a doctor, a family or general practitioner, who has just seen a patient with symptoms of bloating, cramps and diarrhoea that suggest a diagnosis of irritable bowel syndrome. In a typical publicly funded clinic she has around 10 minutes to come to this conclusion. After the consultation and following the doctrine of evidence-based medicine she decides to look up the latest research into the condition. The place to look is the published, peer-reviewed literature, and the most straightforward way to do so is to use PubMed, rather like a medical Google. A quick search on the words 'irritable bowel syndrome' produces thousands of hits. Adding the word 'trial' reduces the number of hits to hundreds and the further addition of 'randomized' brings it down to tens. This is still far too many papers to read. A quick trawl of the titles is bewildering: 'A randomized, controlled exploratory study of clonidine in diarrhea-predominant irritable bowel syndrome' or 'Antidepressants in IBS: are we deluding ourselves?' or even 'Treatment of irritable bowel syndrome with herbal preparations: results of a double-blind, randomized, placebo-controlled, multi-centre trial'.

And PubMed is not the only place to search for information. There is a register of current and recently completed clinical trials in the USA alone that produces yet more hundreds of hits on irritable bowel syndrome. The conscientious doctor is staggering under the weight of information – it would be totally impossible for her to follow up all these leads. Consider, too, that this is just one of her patients. Every single patient has the potential to call up an equally large amount of data that, in theory, must be trawled through to live up to the aspirations of evidence-based medicine.

This information overload is mirrored by anyone seeking something from the Internet. Much is made of today's 'information society'. Vast quantities of facts, fiction, opinion and frank lunacy are available on the World Wide Web, and debate rages over whether or not this increases our ability to understand ourselves and our world. Some argue that the more information available, the more informed the human race becomes; others urge that raw information is useless unless analyzed and validated.

The situation we have today has both of these ideas running in parallel. Huge amounts of raw data are published on the Internet for all to see – you can go and read the entire three billion letters of the human genome if you so wish. Nevertheless, some of the most heavily trafficked web sites are those that practice traditional journalism. The two most popular classes of web site – after pornography – are search engines and news sites. Therefore the solution to information overload on the Internet is to get someone to read it for you and produce a précis, which is essentially what good journalism should do. The solution to information overload in medicine is broadly similar: clinical trials are brought together and assessed in what are known as systematic reviews.

One of the first systematic reviews was 'A treatise of the scurvy. In three parts. Containing an inquiry into the nature, causes and cure, of that disease. Together with a critical and chronological view of what has been published on the subject'. It was published in Edinburgh by James Lind in 1753. Lind, the son of a Scots merchant, joined the navy and was promoted to Ship's Surgeon aboard the 50-gun destroyer *HMS Salisbury*. During a voyage in 1747 Lind conducted a clinical trial, though it wasn't called that at the time, on six different treatments for scurvy, the Vitamin C deficiency endemic in the long-distance sailors of the day. The treatments he assessed were cider, elixir of vitriol (a mixture of sulphuric acid, ginger and alcohol), vinegar, sea water, oranges and lemons, and a purgative. He concluded that oranges and lemons were best.

A year later he retired from the Navy and settled down to practice medicine in his native Edinburgh. Still interested in scurvy he gathered up the few other studies that had been done at the time and brought them together in his treatise. This is the essence of a systematic review. Data from clinical trials are considered side by side and a reasonable conclusion is drawn from them if at all possible. Just as journalism is a digest of current affairs evidence, so a systematic review is a digest of clinical evidence. There is one important difference: systematic reviews are conducted transparently. Decisions about which information is included and the significance it is given are set out in the review.

Lind's research eventually led to ships carrying stores of citrus fruits on long journeys and deaths due to scurvy dwindled to virtually zero. Tellingly, it took nearly 50 years from the publication of his first book for Lind's ideas to become universally accepted. Part of the *raison d'être* of evidence-based medicine is to translate proof of principle as quickly as possible into practice.

After Lind's work there were a few other references to the need for data round-ups to draw useful conclusions. A gentleman farmer, Arthur Young, wrote in 1770 that it was impossible to come to a decision about the relative merits of different agricultural methods if they had just been tested with single experiments on different pieces of land. A century later, the Cambridge physicist Lord Rayleigh berated scientists for not recognizing that data gathered by research needs to be evaluated, not just collected.

Systematic reviews are detailed pieces of analysis and require considerable technical skill to produce. The first stage in the process involves searching the medical literature for suitable clinical trials to include in the review. Today this involves trawling electronic databases of publications and identifying any trial that might be of value to the question the review is seeking to answer. The next stage is to make a detailed examination of the relevant trials to assess their validity. This includes taking careful note of how many patients were involved. Was it double blind, single blind or open? Was it placebo-controlled or a comparison of different treatments? How marked were the effects? How reliable were the statistics? Each trial is then given a weighting, and some with no relevance or poor quality data or bad experimental design are discarded completely. More weight, for example, might be given to the results of a trial on 1,000 individuals with double blind controls than one on 10 patients in with no controls. Putting together a systematic review is not for the fainthearted: the Cochrane Collaboration how-to handbook for reviewers runs to 256 pages.

Reviewers must have a profound understanding of clinical trial methodology and considerable statistical expertise. Most doctors are not particularly well versed in either. The person credited with doing most to advance systematic reviews is the epidemiologist Archie Cochrane (1909–1988). Born in Scotland and educated at Cambridge and London Universities, Cochrane went to fight in the International Brigade in the Spanish Civil War when newly qualified in medicine. Following that he served in the Royal Army Medical Corps in the Second World War, during which he was captured and spent time as a prisoner. After 1945 he did a postgraduate epidemiological study of tuberculosis in the USA and spent the rest of his long career in Cardiff, Wales.

In 1972 Cochrane published *Effectiveness and Efficiency: Random Reflections on Health Services*, now considered the original textbook on evidence-based medicine. In it he outlined his ideas on the importance of doing properly controlled randomized trials and then integrating the results into systematic reviews. Shortly afterwards Cochrane's ideas bore fruit in the form of a series of trials and reviews of medicine as practised on newborns. This led to the establishment of the National Perinatal Epidemiology Unit, based in Oxford in the UK and funded by the World Health Organization and the UK Government. In 1985 a team of 50 volunteers, led by Cardiff obstetrician Iain Chalmers, who knew Cochrane and was impressed by his ideas, published a bibliography of 3,500 reports of controlled trials in perinatal medicine. It was a mammoth effort and a convincing proof of principle. At the same time Chalmers established an international collaboration to evaluate health care in newborns. Seven years later the Director of Research and Development in the British National Health Service, approved funding for a research centre 'to facilitate the preparation of systematic reviews of randomized controlled trials of health care' and the first Cochrane Centre opened in Oxford. A year later Iain Chalmers and about 70 other researchers around the world launched the Cochrane Collaboration, 'to prepare, maintain, and disseminate systematic reviews of the effects of health care interventions'. Today this international effort is recognized as the world leader in evidence-based medicine.

At the heart of the Collaboration's work are the Cochrane Reviews. These substantial documents are published quarterly in print and online. They cost between £150 and £350 to access in full, but synopses are available for all to search on the web. Each review provides a snapshot of the latest research into a particular treatment for a specific condition.

This is the answer that has emerged to the problem of information overload. Our imaginary doctor vainly trying to establish the best treatment for irritable bowel syndrome would, in practise, probably do a quick search of Cochrane Reviews. This turns up fewer than 10 hits for the condition, each a paragraph summary. So, with one minute's searching and five minutes' reading, the doctor can be reasonably sure of tapping into most of the latest information on the topic. Cochrane Reviews do not cover every subject, but the library is constantly being augmented and each review is updated on a regular basis, normally every two years or so. If a Cochrane Review exists it is seen as one of the most reliable sources of evidence-based medicine currently possible.

Fifty international expert groups currently contribute to the Cochrane Library. Together they produce hundreds of reviews each year, coordinated by the group editorial teams. As with all scientific publications these reviews are themselves peer-reviewed before acceptance and there is a formal mechanism by which interested researchers can comment on their conclusions.

There are other sources of systematic reviews, though few as comprehensive as the Cochrane Library. Web journal *Bandolier*, run by an independent group of scientists based in Oxford, England, is one widely respected fount of evidence based wisdom. Another is *Clinical Evidence* from the BMJ Publishing group. And increasingly medical journals such as the *New England Journal of Medicine*, *The Lancet*, the *Journal of the American Medical Association* and the *British Medical Journal* are publishing regular systematic reviews. The evidence base for evidence-based medicine is mushrooming.

But evidence-based medicine, like clinical trials, has nothing to say about how – physically, chemically, biologically – a particular treatment works. It simply focuses on whether and to what extent it is effective and in what individual patient circumstances. As such it is blind to many of arguments surrounding complementary and alternative medicines. If a therapy works – as assessed by clinical trials – from surgery to shiatsu, evidence-based medicine will welcome it into the fold; if it does not it will be rejected.

Many complementary researchers and practitioners are highly sceptical of evidence-based medicine. They feel that it is medicine by rote. Here's a patient showing symptoms x, y and z; enter these into the evidence-based flow chart and out will come the perfect treatment. Evidence-based medicine fans reject this criticism as a caricature, but the idea persists – partly because of the gulf in perspective, or at least the perceived gulf, between complementary practitioners and medical orthodoxy. To most complementary therapists the relationship with the patient is a highly interactive, personal and essential part of the healing process. The training and practice of many complementary therapies emphasizes that no two patients are the same and great weight is laid on the practitioner's intuition. Faced with a treatment regime – evidence-based medicine – that appears to emphasize external information sources over the therapist's expertise, complementary practitioners are often instinctively hostile.

A good way to understand what evidence-based medicine means to doctors and patients is to look at how it is taught. One of the standard evidence-based medicine textbooks for doctors, *Evidence-Based Medicine – How to Practice and Teach EBM*, edited by David Sackett, breaks the process down into five steps.

Step one involves forming answerable questions to the problems the patient is presenting. A question such as 'Are antibiotics the best treatment for my 87 year old female patient's sore throat?' is much easier to answer than 'What is wrong with this patient?'. Detectives, investigative accountants and management consultants take a similar approach.

Step two is to track down the best evidence to answer step one. Doctors are urged to search through medical journals, textbooks and, increasingly, online databases.

Step three is to critically appraise the evidence information collected. Is it relevant to the condition the patient is presenting? Is the evidence applicable to that problem and how big an impact does it have?

Step four is to combine the relevant evidence as sifted out in step three with the doctors' own clinical expertise and the individual biology and circumstances of the patient.

Finally, step five: self-appraisal. Doctors should, according to this rubric, reflect on how well they performed steps one to four and how they might do better in future.

A doctor following these five steps, the reasoning goes, will give his or her patient the best therapy currently available, budget permitting. This is a bold claim, and no evidence-based medicine advocate argues that it is achieved all the time. For example, the same textbook accepts that most practising doctors act only on stage two, assessing the evidence, most of the time.

In fact, the differences between the practice of evidence-based medicine and complementary therapies are not as great as they might appear. At best, both blend general principles with the individual patient's circumstances and the clinical expertise of the doctor or therapist. Both seek the best possible outcome from the available data and both seek to improve as more information becomes available. For doctors that might be another systematic review and for an alternative therapist it may be another practitioner's intuition, but the aim is the same.

○

Even the most ardent supporters of evidence-based medicine accept that it has its drawbacks. Many of these are described in a paper by William Rosenberg and Anna Donald published in April 1995 in *The British Medical Journal*. Evidence-based medicine, they point out, takes time to learn and to practise. As a result, senior doctors running clinics need to develop good management skills to ensure that juniors in their charge have the time to acquire and apply the evidence. Doctors need to become skilled at searching computer databases, which can be a challenge for older practitioners less familiar with the technology. Gaps in evidence, while important for driving future research, need skill to navigate. Finally, suggest Rosenberg and Donald, authoritarian clinicians can see evidence-based medicine as a threat. 'It may cause them to lose face by sometimes exposing their current practice as obsolete or occasionally even dangerous.' That anyone with an Internet connection and some understanding of the terminology can access the same information as the most senior hospital specialist has the potential to demystify the practice of medicine. It's tantalizingly egalitarian.

Small wonder, then, that evidence-based medicine has been given a mixed reception by those who have to put it into practice: doctors. A telling study of 15 Canadian family physicians (GPs) was published in May 2003 in the journal *BMC Family*

Practice. Using semi-structured interviews the authors elicited these physicians' multi-faceted views of evidence-based medicine, then developed a qualitative analysis of those interviews.

The family physicians gave a guarded welcome to evidence-based medicine, admitting that it had improved doctor–patient communication and standards of care, but they were worried about several other consequences. Its lack of emphasis on intuition downplayed something they valued highly. The doctors also felt that evidence-based medicine devalued 'creative problem solving in family practice' and 'the art of medicine' and that guidelines were 'a constraining force on family physicians'. All 15 reported having run into conflicts as they attempted to put evidence-based medicine into practice. Patients' preferences could often be at odds with the evidence, for example. As one physician observed: 'Sometimes it's hard to sell it [the evidence] to certain patients. They have a certain expectation and family medicine is to be patient-centred'. Another admitted that although they explained the evidence to patients to the best of their ability 'we end up doing what the patient wants most of the time'. The researchers conclude by applauding any revisions to the practice of evidence-based medicine that places a greater emphasis on clinical expertise and patient preferences. This, they had shown 'can serve as trumps to research evidence'. (Ironically, these are the very themes that complementary therapists set such store by.)

This chimes with the experience of general practitioner Graham Ward, a doctor of 15 years standing who works in a mixed practice in Bristol in the West of England. He agrees that evidence is an important part of his practice but that it can be a hindrance as well as a help.

Ward worries that some elements of evidence-based medicine have been seized on by governments as a way of measuring doctor performance. This has become particularly significant in the UK, where the incumbent administration has imposed targets for many publicly funded groups, including doctors. Ward says that he has 'certain boxes to tick' founded on an evidence-based approach and gave me an example he had experienced that day.

A patient had come to Ward's clinic complaining of agonizing headaches. On inspection of their notes Ward found mention of an epileptic fit the patient had experienced 10 years previously. The patient had not suffered one since and, according to Ward's judgement, this was not likely to be related to the headaches that were currently being experienced. Nevertheless, he had to question the patient about the epileptic episode and noted that he had done so. It was, he said, probably totally irrelevant to the patient's condition and nothing to do with the reason they were sitting in his surgery crying with headache pain. Yet it was necessary for Ward to include a discussion on epilepsy to, as he put it, 'satisfy our ringmasters'.

Graham Ward doesn't reject evidence-based medicine. Far from it: he requires a good level of evidence of the effectiveness of a drug before he will prescribe it and is cautious about welcoming complementary therapies. What he expresses is pragmatism. As he is fond of saying, 'there is evidence-based medicine and there's common-sense medicine'. Yet his objections are interesting. Is box-ticking a waste of time? Could what appears to be an irrelevance be an unkind intrusion? Or is it a small price to pay for ensuring that doctors do their work carefully and above all, systematically?

Evidence-based medicine has had a similarly mixed reception amongst complementary therapists. Some view it as yet another mechanistic reductionist approach and therefore unsuited to the patient-oriented, individualistic focus of alternative medicine. Others believe it has the potential to further the integration of complementary and orthodox approaches.

Paul Dieppe, for instance, is sceptical about applying biomedical research techniques to complementary medicine.

The point he makes about clinical trials being poor at measuring individualized treatments has a knock-on effect for evidence-based medicine. If the evidence is not available, he argues, there can be no evidence-based medicine. Inevitably, complementary therapies are going to be under-represented.

○

Evidence-based health care draws upon the same type of data as its medical twin, so the arguments about what counts as evidence and how to gather it are equally applicable. The different is that evidence-based health care guides policy, not individual treatments. Crudely put, it is a way of spending limited funds to achieve maximum benefits.

Broadly there are two models of health care funding: payment at the point of delivery – private medical care – or contribution to a fund that pays out when required. That fund might be a private medical insurance company, a government-sponsored insurance scheme or a state-run health care system funded by taxation. In each case, there is considerable pressure on those responsible for spending to do so wisely and to be accountable for their decisions.

BUPA is a health and care company which began in 1947. It has grown into an international organization with a yearly income of almost £3.5 billion and more than seven and a half million customers in more than 180 countries.

BUPA UK Membership is responsible for deciding which clinicians are eligible to treat UK members. At present that includes four complementary therapies: osteopathy, chiropractic, acupuncture and homeopathy. BUPA UK Membership receives regular requests for others to be added to the list. Issues that need to be explored when reviewing treatments that may be eligible for reimbursement include clinical evidence, cost effectiveness and member demand. The evidence is assessed by a medical team drawing on the familiar sources of systematic reviews of clinical evidence and trials.

BUPA is constantly reviewing the services it offers and the treatments it funds and an evidence base is central to this process. The organization does not fund any research into complementary medicine from its own resources, but that is not necessarily the case elsewhere.

In Germany, public health cover is managed by a group of insurance companies. Until a few years ago a loophole in the system allowed individuals to claim for some one-off treatments, including some complementary therapies. That loophole has now been closed, resulting in many patients asking for alternative medicine to be included under the scheme.

At the time of writing, a series of clinical trials for acupuncture to treat migraine, osteoarthritis and backache are in progress, instigated and paid for by German insurers, in response to consumer demand. If the results are positive then it is likely that acupuncture for these conditions will be made available within the German health care system.

○

One of the undercurrents of this book has been something that is really a political (with a small 'p') question. Namely: what is acceptable medicine – no matter that it is called orthodox, regular or conventional – and what is not acceptable medicine, regardless of whether it is labelled charlatanism, quackery or unorthodox? It is a question about boundaries between the two. Are they shifting, should they be retained and how they should be 'policed'? It is also a question about protecting a vulnerable, credulous public from, at worst, the blandishments of the next Lydia Pinkham (a 19th century Massachusetts woman who produced an alcohol-based 'Vegetable Compound' for the treatment of menstrual pain) peddling trumped up 'wonder cures'.

Conventional medicine is currently acceptable medicine. Perhaps its key feature is that it is *scientific*. That is, it is medicine that is based on predominantly western thought of the past 200 or so years, as distinct from folklore or ancient tradition. It tends to rule out non-western systems of medical practice, such as Ayurveda or Traditional Chinese Medicine.

Western science is based on a particular idea about experimental proof – involving a specific set of theories of knowledge. Evidence counts if it has been collected as a result of a well-designed, properly controlled experiment, the results of which have been published in peer-reviewed journals for other scientists to comment upon. Now, though, research into complementary medicine is not only questioning the way evidence is gathered but also challenging the notion of what counts as evidence.

Social anthropologist Christine Barry raised the question of what is to count as evidence in a paper presented to a conference in July 2003. She took a look at much of what this book has covered, but came at it from a rather different angle. Social anthropology, she argued, is very familiar with many of the ideas that makes complementary medicine distinct.

On the one hand, complementary medicine entails holism, the emphasis on the social interaction between practitioner and client as integral to the therapy, and on the patient's ideas about the meaning of their bodily experience. On the other hand, concepts such as 'transcendent, transformational experiences', 'giving meaning', 'changing lived-body experience' or 'intersubjective consensus' are readily understood in anthropology and there is a tradition within that discipline of offering evidence based on these ideas. What's more, she points out, social anthropology has developed the tools to examine these concepts – ideas that biomedicine largely ignores.

Barry argues that social anthropology has the potential to produce analyses that *can* be counted as evidence about complementary medicine. Anthropology's theory of knowledge – including definitions as to what evidence can look like – is very different and might, therefore, suit complementary medicine better than the theories associated with scientific medicine.

Combining the two will be a big challenge – social anthropologists and clinical researchers speak very different languages. But in the opinion of Barry and others like her, it is a must: the current definition of what counts as evidence is, they say, too narrow. The only way to get a complete description of complementary medicine is to expand the definition of evidence beyond its narrow biomedical confines.

○

The issue of what constitutes evidence is not unique to complementary medicine, as one brief final example suggests. In 1999 the UK government published a White Paper outlining its policy on public health, entitled 'Saving Lives: Our Healthier Nation'. It contained the following paragraph:

> Research plays a major role in helping us understand better the causes of ill health.... Public health research is also important in establishing the effectiveness of health programmes but we need to widen the scope of the methods used beyond the randomised controlled trial. In the past it has been the gold standard for research but it is no

> longer applicable to all the kinds of research questions which need to be answered.

The sort of public health programmes this paper referred to include opening new hospitals or urban regeneration. Implemented in schools, neighbourhoods or workplaces, these large-scale projects are inseparable from the social interactions. Health and safety at work, for example, rely on employers and employees cooperating. Traffic calming schemes need the support of the local community. These programmes take years to implement and even longer to see results from.

This is all sounding very familiar: randomized controlled trials are not applicable, and there are long-term effects, social interactions and complex interventions. The challenges facing public health planners have many parallels with those facing complementary and alternative medicine researchers. Once again, attempts to get a handle on complementary medicine have potential applications further afield.

10

conclusion

No single technique is emerging to answer the question: 'Are complementary therapies effective?'. What's more, if answers are to come they will likely do so from a combination of different approaches. There is still a debate around the questions 'What does effective mean?' and 'What should be measured and with what technique?'. Research on placebo effects is muddying the water of whether a placebo-controlled trial can provide clear answers, and anyway many of the therapies under discussion do not yet have a suitable placebo. Comparing complementary and orthodox therapies is another option, but they may be working in different ways towards different outcomes – allopathic medicine aiming for a cure, complementary medicine for well-being.

The problems of researching complementary therapies are neither new nor unique. Designing good clinical trials has always been difficult; if it were straightforward the medical literature would not be peppered with arguments over methodology and interpretation of results.

The complexity of the doctor–patient relationship is a challenge to scientific method that is best at examining one element of a system whilst keeping others constant. Yet science deals with complex systems daily.

The big surprise of the Human Genome Project was that we have only around 33,000 genes – way below the 100,000 that many predicted were needed to describe a human being. This dramatically confirmed that the way in which genes work has to be more complex than one gene controlling one function.

In fact, the modern view of how the information in genes controls growth and development is fantastically complex. Switching on a single gene might require anything up to 10 other genes to be active at the same time. And the result of turning on that single gene might influence a whole range of others, turning some on and some off, or changing the rate at which they are read. If one gene has one function, 33,000 genes can only perform 33,000 functions. If one gene works in concert with one or more other genes then the number of possible outcomes rapidly gets into billions.

Geneticists do not, though, throw up their hands and bleat that it is all too hard to study. Instead they design experiments with as few variables as possible, ideally just one, and see what happens. Then they stand back and attempt to fit their new morsel of data into the bigger, more complex, picture. In effect, they switch between viewing the cell as a whole and studying its components in as much isolation as possible. This is, perhaps, not ideal and there will inevitably be some incorrect assumptions made, but it is a way of applying a reductionist method to a complex system.

There are plenty of other examples of this big picture/little picture tactic. Meteorologists do not study the entire global weather system in one go: they break it down into manageable chunks. Ecologists do not assess the impact of global climate change on every shrew and blade of grass: they make generalizations.

It is too glib to say these approaches might be directly applicable to the study of complex interventions in medicine, but lessons could – should – be learnt from studying how science handles complexity in other situations.

○

The whole debate about complementary medicine throws into relief a much bigger question about the medicine we use in general. Many things that doctors currently give their patients have not been through clinical trials. The antibiotic ciprofloxacine has never been tested in children because in animal studies it appeared to cause cartilage damage in juveniles. Nevertheless, it is used with no ill effect when children have an infection that cannot be tackled by another antibiotic. All the empirical – case by case – evidence is that ciprofloxacine is safe; still, this usage has not formally been licensed. Likewise, the treatment of back pain with low doses of the anti-depressant Amitriptyline is highly unlikely to cause a problem, but any doctor prescribing the drug for back pain rather than depression takes a risk. If something goes wrong he does not have any legal support for prescribing 'off licence'.

This is not to say that tests on complementary medicines should be waived. Rather, it means that rejecting complementary therapies purely on the grounds that they have not been fully tested is inconsistent.

○

Chronic conditions and slow degenerative diseases are on the increase. With no cures in sight, palliative care is becoming a bigger part of medicine. Complementary practitioners appear to be comfortable with helping people live with disease rather than attempting to cure it. Many I spoke to talked of the pleasure they get from enabling someone to get on with their lives a little better even though they both know that the problem will never go away. It may be that this success is no more or less than a placebo or context effect. Nevertheless, evidence from psychoneuroimmunology suggests a very real connection between the mind and the body through which these effects might work.

For people like me, with a grounding in biological science, the demonstration of a biochemical link is reassuring. It may have no impact at all on how effective a placebo or context might be, but it fits more comfortably into the biomedical model. And yet it is important to remember that the usefulness of a treatment is totally independent of how it works. Complementary medicine's apparent ability to elicit a powerful placebo effect is the point, rather than how it does it.

That said, access to any sort of health care provision is becoming increasingly evidence-based. If complementary medicine is to feature in the publicly or insurance company-funded system – and there clearly is a demand for it – then it will come under even closer scrutiny. The challenge for complementary advocates is, perhaps, to convince those who hold the purse strings that evidence can come from a wider range of sources than a narrowly defined randomized controlled trial.

This raises a particular paradox for those calling for a plurality in health care, arguing that the complementary approach should live alongside orthodox medicine without needing to conform to its measures of success. Currently, true plurality is only available to those who can afford to pay.

There is a case to be made that randomized controlled trials based on biomedical outcomes do not capture all the elements of many complementary therapies. Adaptations of those trials might get closer to a fuller explanation of what is going on. Qualitative research may well help to develop those tweaks and provide valuable insights unconnected to the effectiveness of a treatment.

There are some intriguing and currently inexplicable findings surrounding complementary therapies. Why, for example, does the Bowen Technique appear to work so well for frozen shoulder? What is the explanation behind the apparent ability of homeopathy to be more than a placebo? These results present some very profound challenges to science's explanation of how our bodies work. This can only be a good thing. The scientific method is a powerful tool for investigating the unknown, and research thrives on puzzles. Attempting to answer these questions will certainly produce new data. It may be as mundane as spotting the hidden flaw in the research, or it may uncover a previously unknown function of our bodies.

Pursuing these puzzles has another potential benefit. Testing complementary medicine is difficult. In science, doing difficult things often results in better tools. Even if every single complementary therapy turns out to have a simple explanation, extracting it is going to hone medical research. That in turn will improve the way in which orthodox medicine can be studied.

It is appropriate to end with a conclusion drawn by someone who has spent his working life as one of the most senior and respected medical researchers, and who has been closely associated with the development of randomized controlled trials and with the establishment of the Cochrane centre:

> the most important resource required to promote the concept of integrated health care is likely to be humility among those whose practices will be put to the test, within both orthodox and complementary medicine.

Professor Sir Iain Chalmers, 1998

further reading

Australian Government Expert Committee on Complementary Medicines in the Health System Report to the Parliamentary Secretary to the Minister for Health and Ageing (2003) *Complementary Medicines in the Australian Health System*, September. Available at http://www.tga.gov.au/docs/html/cmreport1.htm.

Bandolier: http://www.jr2.ox.ac.uk/bandolier/.

Barnes, P., Powell-Griner, E., McFann, K. and Nahin, R. (2004) *Complementary and Alternative Medicine Use Among Adults: United States, 2002*. CDC Advance Data Report #343, 27 May.

Berman, J.D. and Straus, S.E. (2004) Implementing a research agenda for complementary and alternative medicine. *Annual Reviews of Medicine*, **55**, 239–54.

BMA News (2004) Complements of the house. London: British Medical Association, 22 May.

British Medical Journal (2004) Clinical Evidence: the international source of the best available evidence for effective health care. British Medical Journal Publishing Group Ltd. Available at http://www.clinicalevidence.com/.

Cochrane Reviews: http://www.cochrane.org/.

Cohen, I.R. (2000) *Tending Adam's Garden*. Academic Press: San Diego, CA.

Coward, R. (1989) *The Whole Truth: the Myth of Alternative Health*. London: Faber.

Department of Complementary Medicine, Peninsula Medical School (2004) *Complementary Medicine: The Evidence So Far*. A documentation of research 1993–2003. A summary of our most important research findings to date. Exeter: Universities of Exeter and Plymouth.

Earl-Slater, A. (2002) *The Handbook of Clinical Trials and Other Research* Abingdon, Oxfordshire: Radcliffe Medical Press.

Evans, D. (2003) *Placebo: the Belief Effect*. London: HarperCollins.

Evans, P., Hucklebridge, F. and Clow, A. (2000) *Mind, Immunity and Health: the Science of Psychoneuroimmunology*. London: Free Association Books.

Freidson, E. (1980) *Patients' Views of Medical Practice: A Study of Subscribers to a Prepaid Medical Plan in the Bronx*. Chicago: University of Chicago Press.

Ga^{2}len (2004) *Global Allergy and Asthma European Network: Network of Excellence*. European Union, 6th Framework Programme. Available at http://www.ga2len.com/.

Global Polio Eradication Initiative (2003) *Strategic Plan 2004–2008*. World Health Organization Publications. Available at http://www.polioeradication.org/.

Gray, A. (ed.) (1993) *World Health and Disease*. Milton Keynes: Open University Press.

Muir Gray, J.A. (2001) *Evidence-Based Healthcare*. London: Churchill Livingstone.

Guess, H.A., Kleinman, A., Kusek, J. and Engel, L. (eds.) (2002) *The Science of the Placebo: Toward an Interdisciplinary Research Agenda*. London: BMJ Books.

House of Lords Select Committee on Science and Technology Sixth Report (2002) *Complementary and Alternative Medicine*. London: The Stationery Office, 21 November. Available at http://www.parliament.the-stationery-office.co.uk/pa/ld199900/ldselect/ldsctech/123/12302.htm.

Illich, I. (1977) *Limits to Medicine: Medical Nemesis – The Expropriation of Health*. London: Penguin Books Ltd.

Jenkins, T., Campbell, A., Cant, S., Hehir, B., Fox, M. and Fitzpatrick, M. (2002) *Alternative Medicine: Should We Swallow It?* Institute of Ideas: Debating Matters. London: Hodder & Stoughton.

Karpf, Anne (1988) *Doctoring the Media: the Reporting of Health and Medicine*. London: Taylor & Francis.

Lewith, G.T. and Aldridge, D. (1993) *Clinical Research Methodology for Complementary Therapies*. Singular Publishing Group.

The James Lind Library: Documenting the evolution of fair tests of medical treatment. Available at http://www.jameslindlibrary.org/.

Mackay, H. and Long, A.F. (2003) The experience and effects of shiatsu: findings from a two country exploratory study. *Report no. 9*. Salford: University of Salford Health Care Practice R&D Unit.

Martin, P. (1998) *The Sickening Mind: Brain, Behaviour, Immunity and Disease*. Flamingo.

Mason, J. (2002) *Qualitative Researching*, 2nd edn. London: Sage.

McKeown, T. (1976) *The Modern Rise of Population*. London: Edward Arnold.

Murphy, E. and Dingwall, R. (2003) *Qualitative Methods and Health Policy Research*. New York: Aldine.

Murphy, E., Dingwall, R., Greatbatch, D., Parker, S. and Watson, P. (1998) Qualitative research methods in health technology

assessment: a review of the literature. *Health Technology Assessment*, **2**(16), 1–276.
Peters, D. (ed.) (2001) *Understanding the Placebo Effect in Complementary Medicine.* London: Churchill Livingstone.
Porter, R. (2002) *Blood and Guts: a Short History of Medicine.* London: Allen Lane.
Sackett, D.L., Straus, S.E., Richardson, W.S., Rosenberg, W. and Haynes, R.B. (2000) *Evidence-Based Medicine: How to Practice and Teach EBM*, 2nd edn. London: Churchill Livingstone.
Seale, C. (2002) *Media and Health.* London: Sage.
Sharma, U. (1992)*Complementary Medicine Today: Practitioner and Patients.* London: Routledge
World Health Organization (2001) *Legal Status of Traditional Medicine and Complementary/Alternative Medicine: A Worldwide Review.* Available at http://www.who.int/medicines/library/trm/who-edm-trm-2001-2/legalstatus.shtml.

Index